S. Brime

Plant Molecular Genetics

Front cover picture: Examples of *P-pr* phenotype of maize. The variegated pigmentation pattern of the maize kernal pericarp shows the timing and frequency of epigenetic switches at the *P*-locus in maize. This gene encodes a transcription factor which controls pigment synthesis in maize. The epigenetic switch reverses modification of the *P*-locus that is stably transmitted as an allele (*P-pr*). The standard, full colour, *P-rr* allele is shown in the top frame. The *P-pr* allele has a colourless phenotype that reverts to a normal red phenotype during plant development. The changes are quite variable among progeny and frequently show ear sectors as well as kernal stripes. These changes of the phenotype are correlated with the demethylation of site-specific hypermethylated sites of the *P* gene (see Chapter 1).

Back cover picture: *Arabidopsis thaliana* showing the phenotype of the *terminal flower* (*tfl*-2) mutation (see Chapter 11). Reproduced by permission of the Nottingham *Arabidopsis* Stock Centre, University Park, Nottingham, NG7 2RD, UK.

Plant Molecular Genetics

Monica A. Hughes

Longman

Addison Wesley Longman Limited
Addison Wesley Longman Limited,
Edinburgh Gate, Harlow,
Essex CM20 2JE, England
and associated companies throughout the world

First Published 1996

British Library Cataloguing in Publication Data
A catalogue record for this title is available from the British Library

ISBN 0-582-24730-6

Library of Congress Cataloging-in-Publication Data
A catalog entry for this title is available from the Library of Congress

Set by 30 in 10/12pt Times
Printed in Great Britain by Henry Ling Ltd., at the Dorset Press,
Dorchester, Dorset.

Contents

Part Four An introduction to plant biotechnology

Preface

In the past 10 years plant science has emerged from the tweed jacket, pressed flower and carrot slice era, to become an exciting, intellectually demanding and rapidly developing subject, which offers careers in new applications of the science. This change in plant science research has come about largely because of the application of molecular genetic techniques to the study of plant processes. This book aims to present higher plant (angiosperm) molecular genetics in a format that is easy to assimilate by undergraduate students. The book is based on an undergraduate course (single module) given to genetics students at Newcastle University. It assumes an A-level standard of knowledge of plant biology and a fundamental or first-year undergraduate knowledge of molecular techniques and classical genetics.

The book is selective and not an encyclopaedic account of the molecular genetics of all areas of plant biology. Topics have been chosen partly because of their intrinsic interest and importance in plant biology and partly because of the state of molecular knowledge in that topic. Emphasis has been given to the use of genetic analysis in combination with molecular techniques. This means that studies which have used either the selection and analysis of mutant plants or transformation technology and analysis of transgenic plants have been given a prominent place in the book and this choice reflects the power of this type of analysis in research of plant processes.

A relatively small number of annotated references are given for each chapter that guides the reader to further reading. Each chapter also has a list of major learning objectives. These are intended to help students to assess their knowledge and understanding of each topic. They are all based on the material of the chapter but may expect the student to integrate the information in a way that does not involve a straight recapitulation of the material in the chapter.

The book is organized so that Part 1 provides information about plant genomes and their inheritance and expression. Part 2 deals with the biology of *Agrobacterium tumefaciens* and the use of this bacterium in plant transformation. This topic is given an early and prominent position in the book because of the importance of this technique in many plant molecular studies. Part 3 covers a number of different topics in plant biology and Part 4 gives an introduction to the application of plant molecular genetics in plant biotechnology. The book ends with a short description of the widespread debate about the possible problems of what has been called 'plant genetic engineering'.

Acknowledgements

I wish to thank all those who helped me by critically reading chapters of the book, namely the members of my laboratory and Professor Nigel Robinson in the Department of Biochemistry and Genetics at Newcastle University and Dr John Jones in the Department of Agricultural Botany at Reading University. I would also like to thank Dr Howard Haysom for his invaluable help in the production of the manuscript.

We are grateful to the following for permission to reproduce copyright material:

BIOS Scientific Publishers for Fig. 1.4 (Simpson, Leader and Brown, 1993); CAB International for Fig. 1.2 (Casey and Davies, 1993) and Fig. 7.4 (Mitten *et al.*, 1990); Faber and Faber Limited and Random House, Inc. for 'Progress?' from W.H. Auden, *Collected Poems.* © 1974 The Estate of W.H. Auden.; Kluwer Academic Publishers for Fig. 2.8 (Goddard *et al.*, 1993), Figs 3.2 and 3.3 (Sugiura, 1992) and Figs 4.5 and 4.6 (Levings and Siedow, 1992); Macmillan Magazines Limited for Fig. 2.6 (Patersen *et al.*, 1988); Oxford University Press for Fig. 9.3 (Mott *et al.*, 1989); Plenum Publishing Corporation for Figs 4.4 and 9.1 (Hanson *et al.*, 1993); Dr J. A. Roberts, on behalf of the Society for Experimental Biology for Fig. 13.7 (Ougham and Howarth, 1988); Springer-Verlag for Fig. 1.7 (Seinmuller, Batschauer and Apel, 1986) and Fig. 2.5 (Galiba *et al.*, 1995).

Whilst every effort has been made to trace the owners of copyright material, in a few cases this has proved impossible, and we take this opportunity to offer our apologies to any copyright holders whose rights we may have unwittingly infringed.

List of Abbreviations

ABA	abscisic acid
ABRE	abscisic acid-resposive element
ARS	autonomously replicating sequences
bZIP	basic leucine zipper
bp	base pair
CaMV	cauliflower mosaic virus
CMS	cytoplasmic male sterility
DAG	1, 2-diacylglycerol
DNase	deoxyribonuclease
DRE	drought-responsive element
ERE	ethylene-responsive element
EV	'empty' vector
FUE	far-upstream element
GUS	β-glucuronidase
HR	hypersensitive response
HSP	heat-shock protein
IEF	isoelectric focusing
IP_3	inositol 1, 4, 5-trisphosphate
IR	inverted repeat
Kb	Kilobases (1000 bases)
LRE	light-responsive element
M_r	relative molecular mass
MCS	multiple cloning site
NUE	near-upstream element
ORF	open reading frame
PAGE	polyacrylamide gel electrophoresis
QTL	quantitative trait loci
RER	rough endoplasmic reticulum
RFLP	restriction fragment length polymorphism
RNase	ribonuclease
Rubisco	ribulose, 1, 5-bisphosphate carboxylase/oxygenase
SAR	scaffold-associated region
SARes	systemic acquired resistance
SLSG	S-locus-specific glycoproteins
SRK	S-related kinase
TCS	transcription start site
TMV	tobacco mosaic virus
UTR	untranslated region
WT	wild type

The symbols for genes are shown in italics an in general non-functional alleles of the gene have a lower-case symbol. The symbol for the protein encoded by the gene uses the same letters and numbers (if present) but is not italicized.

Progress?

Sessile, unseeing,
the Plant is wholly content
with the Adjacent.

Mobilised, sighted,
the Beast can tell Here from There
and Now from Not-Yet.

Talkative, anxious
Man can picture the Absent
and Non-Existent.

[From *Collected Poems* by W.H.Auden, published by Faber & Faber Ltd.]

Part one

Plant genome structure

Chapter 1

Nuclear genome

1.1 Plant cells contain separate genomes in the nuclei, chloroplasts and mitochondria

Plant cells contain genetic information within three different organelles. In all of these the genetic information is encoded within large molecules of DNA, called chromosomes, but the structure of the chromosomes varies between the organelles. The nuclei of plant cells contain various linear DNA molecules, the number and length of which vary between different plant species (Table 1.1). The chloroplasts and the mitochondria also contain DNA but in the form of circular molecules. Although individual chloroplasts and mitochondria contain more than one chromosome, within a single organelle each circular chromosome contains the same genes (see Chapters 3 and 4).

Most of the plant's genetic information is contained within the linear chromosomes of the nucleus and, since most higher plants (angiosperms) are diploid, each nucleus contains two sets of chromosomes with each set composed of the same genes. This set of genetic information is known as the nuclear genome. The nuclear genomes of different plant species contain different amounts of DNA. Table 1.1 gives the genome size of some of the plant species referred to in this text. A number of plant species are polyploid, that is the nuclei contain more than two sets of chromosomes. In Table 1.1, barley is shown as a diploid cereal whereas wheat is hexaploid, having six sets of chromosomes.

The genome size is not strictly related to the number of chromosomes: wheat has a genome of seven chromosomes but has approximately 10 times the quantity of DNA present in the potato genome, composed of 12 chromosomes. The wide range of genome sizes illustrated in Table 1.1 is a feature of higher plants and variation may even exist between species in the same genus. *Vicia faba*, for example, has nearly 10 times the amount of nuclear DNA compared with the closely related species *Vicia sativa*, even though these two species have the same number of chromosomes (Table 1.1). In general plant nuclear genomes are large: broad bean has a genome four times the size of the human genome (3.3×10^9 bp) and *Arabidopsis*, a model species for many plant molecular studies chosen partly because of its small genome, has a genome nearly five times the size of the *Drosophila* genome.

Table 1.1 Plant nuclear genome sizes (in numbers of base pairs (bp) in the unreplicated genome; 1C value)

Species	Common name	Monocot (M) Dicot (D)	No. of chromo- somes (1n)	Ploidy level	Genome size (bp)
Antirrhinum majus L.	Snapdragon	D	8	2	1.54×10^9
Arabidopsis thaliana L.	Thale cress	D	5	2	1.00×10^8
Brassica napus L.	Oilseed rape	D	19	2	1.23×10^9
Brassica oleracea	Cabbage	D	9	2	6.0×10^8
Hordeum vulgare L.	Barley	M	7	2	4.8×10^9
Lycopersicon esculentum Miller	Tomato	D	12	2	1.0×10^9
Manihot esculenta Crantz	Cassava	D	16	2	7.5×10^8
Nicotiana tabacum L.	Tobacco	D	12	4	4.4×10^9
Oryza sativa L.	Rice	M	12	2	4.2×10^8
Petunia hybrida	Petunia	D	7	2	1.27×10^9
Pisum sativum L.	Garden pea	D	7	2	4.1×10^9
Secale cereale L.	Rye	M	7	2	9.5×10^9
Solanum tuberosum L.	Potato	D	12	4	1.8×10^9
Triticum aestivum	Wheat	M	7	6	1.6×10^{10}
Vicia faba L.	Broad bean	D	6	2	1.2×10^{10}
Vicia sativa L.	Common vetch	D	6	2	1.6×10^9
Zea mays L.	Maize	M	10	2	2.5×10^9

There is a relationship between the genome size and the life cycle of a plant species, such that annual plants, particularly self-pollinating ephemeral annuals, have small genomes whereas perennial plants tend to have larger genomes. This in turn is related to the relatively faster cell cycle in the annual species compared with perennials and may reflect the constraint on cell cycle time and hence the time for plant development caused by the time required for the replication of large genomes.

Box 1.1 Flax genotrophs

Heritable changes can be induced in certain flax varieties when they are grown in particular environments for a single generation. A typical 'plastic' variety is Stormont Cirrus. When it is grown in 1.5% (mass/volume) ammonium sulphate plus 1.5% (m/v) potassium chloride the plants of the next generation are bigger (height and weight) and the nuclear DNA content is higher than in the original parent plants. These characters are stably inherited even by plants not growing in the 1.5% (m/v) ammonium sulphate plus 1.5% (m/v) potassium chloride nutrient treatment. These stable lines have been called genotrophs. Although DNA amplification has been described in some animal systems, the mechanism of these genome changes in flax is not known.

1.2   Classes of DNA in the nuclear genome

Double-stranded nuclear DNA can be denatured into two single strands by heating. Under suitable conditions, the hydrogen bonding between complementary single-stranded DNA molecules will reform to reanneal the two strands. If all other parameters are kept constant, the rate of reannealing will depend on the concentration of complementary sequences. This means that if a particular sequence occurs a large number of times within a genome this DNA will reanneal faster than sequences that are only present once.

Analysis of reannealing kinetics of plant nuclear DNA shows that much of the genome consists of repeated sequences. These are often arbitrarily divided into: (i) low to moderately repeated sequences and (ii) highly repeated sequences. In most plants only 20–40% of the genome consists of unique or single-copy DNA sequences. The larger plant genomes are associated with more repetitive DNA: pea (4.1×10^9 bp) has been estimated to have 70% repetitive DNA whereas mung bean (*Vigna radiata*), which is in the same family as pea (Leguminosae), has a small genome (5.0×10^8 bp) and only 40% repetitive DNA. Similarly the barley genome consists of about 80% repetitive DNA whereas the small genome of the cereal rice has only 50% repetitive sequences.

1. Low and moderately repetitive DNA consists of sequences repeated from several to thousands of times and is often interspersed with single-copy sequences along the chromosomes. Some of this type of DNA has a known function; thus both histone genes and ribosomal RNA (rRNA) genes are present as multiple copies in plant nuclear genomes. The nuclear genome of *V. faba* has about 4750 rRNA genes whereas the smaller genome of *V. sativa* has only 1875.

2. Highly repetitive DNA is usually short sequences present at 10^5–10^7 copies per genome. This type of DNA is often associated with structural features of the chromosome, such as the centromere (see section 1.3, Centromere). The large blocks of highly repetitive DNA sequences may have a different pattern of condensation during the cell cycle compared with the rest of the genome and this results in a different pattern of staining for these regions of the chromosome, which are termed heterochromatic.

Box 1.2 B chromosomes

In addition to the large chromosomes that show regular segregation during mitotic and meiotic division of the nucleus (see Chapter 2) many plants, including rye and maize, contain small auxiliary or B chromosomes. The segregation of these B chromosomes is random during nuclear division and individual plants will have varying numbers. These chromosomes are largely heterochromatic and their function is not understood. They do not appear to contain the same type of genetic information present in the standard A chromosomes but physiological effects, such as superior fitness of seedlings under conditions of stress in chives and rye, have been associated with B-chromosome number.

1.3 Chromatin structure and chromosome architecture

Box 1.3 Chromosome packing

It has been estimated that a barley diploid nucleus contains 2 m of DNA packaged into a nucleus approximately 10 μm in diameter.

Most current models of the eukaryote nucleus describe a highly ordered structure with individual chromosomes occupying distinct locations within the nuclear matrix. Plant chromosomes have been shown to be attached via their ends (telomeres) and centromeres to the nuclear envelope. There is also evidence from *in situ* hybridization and confocal microscopy that there is clustering of the position of both telomeres and centromeres but there appears to be no general rule and there are insufficient data at present to understand the significance of this clustering.

The long linear DNA molecule of each nuclear chromosome is associated with a group of proteins forming a structure known as chromatin. Twenty years ago the first-order structure of chromatin was resolved and shown to consist of a complex of DNA and five classes of basic proteins called histones (Figure 1.1). Two molecules of each of the histones, H2A, H2B, H3 and H4, make up a core around which two turns (140 bp) of DNA are wound. This structure is known as a nucleosome and these nucleosomes are separated by about 60 bp of spacer DNA, to which one molecule of histone H1 is attached. This coiling reduces the length of the DNA structure to about one-seventh the length of the naked DNA molecule. Acidic proteins (nucleoplasmin and N1) have also been implicated in the nucleosome structure. Histone genes are duplicated in plant genomes and although this is a highly conserved group of eukaryotic proteins variation between different histone genes within a single plant (e.g. wheat) has been found. However, the pattern of variants of H2A and H2B was found to be the same in different developmental stages of wheat, and the significance of histone variants is not known.

The nucleosome string is further coiled to produce a solenoid fibre 30 nm in diameter (Figure 1.1) with the H1 molecules arranged in the centre of the solenoid and the nucleosomes radiating outwards. Intensively transcribed DNA, such as rDNA, is often organized as extended stretches lacking nucleosomes. However, the average gene undergoing transcription is covered in nucleosomes and the relationship between these structures and the expression of genes is not yet known.

Box 1.4 DNase I hypersensitivity

The sensitivity of chromatin to DNase I (endonuclease) digestion (DNase I hypersensitivity) is interpreted as a change in structure associated with a relaxation of the DNA–protein interactions. This is thought to accompany the availability of DNA for transcription and the DNase I sensitivity of chromatin is therefore a feature of genes that are being expressed (see section 1.6).

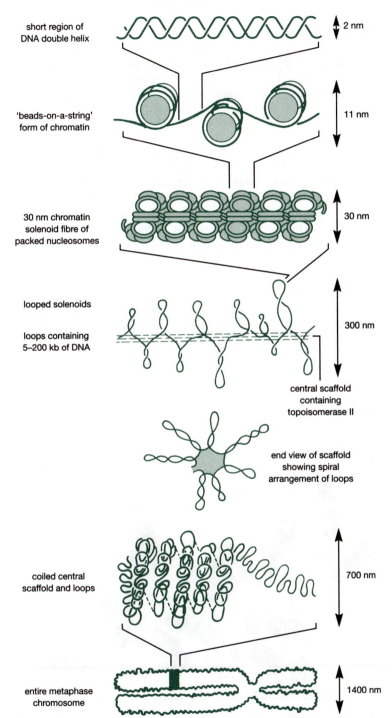

short region of
DNA double helix

2 nm

'beads-on-a-string'
form of chromatin

11 nm

30 nm chromatin
solenoid fibre of
packed nucleosomes

30 nm

looped solenoids

loops containing
5–200 kb of DNA

300 nm

central scaffold
containing
topoisomerase II

end view of scaffold
showing spiral
arrangement of loops

coiled central
scaffold and loops

700 nm

entire metaphase
chromosome

1400 nm

Figure 1.1 Structure of chromatin and the packing of chromatin to form the structure of a mitotic metaphase chromosome.

Box 1.5 Histone 1 phosphorylation

Histone H1 is phosphorylated during the cell cycle, with two phosphate groups added during S phase (DNA replication) and four more added during mitosis. The role of this modification of the protein is unknown. It appears to have no direct effect on chromatin structure but is thought to be involved in chromosome condensation during mitosis.

A further order of organization of chromatin involves the attachment of the 30-nm solenoid fibre at discrete points to a central protein scaffold (Figure 1.1) to produce looped domains containing between 5 and 200 kb of DNA. This structure is approximately 300 nm across and will be further coiled during chromosome condensation in prophase to produce the visible metaphase chromosomes of mitosis (Figure 1.2). The central scaffold is made up of a group of non-histone proteins of which topoisomerase II is a major component.

Box 1.6 Topoisomerase II

Topoisomerase II can catalyse double-stranded breaks in the phosphodiester DNA backbone and is therefore possibly involved in altering the super-helical structure of the domain loops.

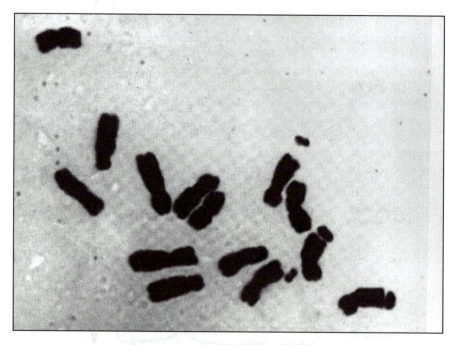

Figure 1.2 Mitotic metaphase chromosomes of pea (Pisum sativum) (2n=14). (From Ellis, T.H.N. The nuclear genome. In Casey, R. and Davies, D.R. (eds) (1993) *Peas: genetics, molecular biology and biotechnology*. CAB International, Wallingford.)

The scaffold-associated regions (SARs) of the DNA contain A/T-rich boxes and sequences homologous to the *Drosophila* topoisomerase II consensus cleavage sequence. The evolutionary conservation of SARs in fungi and animals suggests that plant chromosomes are organized in the same way and a yeast SAR (ARS-1) can bind to plant nuclear protein scaffolds *in vitro*. Recently a number of observations suggest that the scaffold attachment points function as domain boundaries, which play a role in regulating gene expression in plants.

The plant nuclear chromosomes are replicated during S phase of the cell cycle and in order to achieve replication of such large molecules in a relatively short time each chromosome is divided into a number of replicons. In *V. faba* there are 35 000 replicons in the genome, each consisting of about 300 kb. Sequences that confer the ability to replicate have been identified in yeast and are called autonomously replicating sequences (ARS). They contain an 11-bp A/T-rich core sequence (Table 1.2).

Centromere

Most eukaryotic chromosomes contain a specialized region known as the centromere. This is often heterochromatic and its position on the chromosome defines the structure of the chromosome, such that metacentric chromosomes have a centromere in the middle and hence two arms of equal length, whereas an acrocentric chromosome has arms of unequal length because the centromere is not centrally positioned. Some chromosomes have a centromere at one end and are known as telocentric chromosomes.

During prophase of nuclear division the centromere acquires a complicated protein structure, known as the kinetochore. This is the site of attachment of the chromosome to the spindle microtubules. Eukaryotic centromeres contain repetitive DNA and in yeast a consensus sequence (CEN) has been identified that is necessary for stable chromosome inheritance (Table 1.2).

Telomere

The ends of the linear nuclear chromosomes are called telomeres. They contain simple short DNA sequence repeats with a highly conserved G-rich consensus sequence (Table 1.2) found in many organisms including higher plants. This repeated DNA sequence is associated with a special replication process catalysed by an enzyme called telomere terminal transferase. This enzyme contains an RNA molecule that is complementary to the DNA strand of the chromosome which contains a 3' end. The specialized form of replication of the ends of linear DNA molecules is necessary because of the requirement for a double-stranded region to prime DNA polymerase and the movement of the DNA polymerase 3' → 5' along the template strand. In addition to the requirement for some specialized replication, the ends of DNA molecules are probably susceptible to exonuclease digestion and the non-replicating telomere is complexed with a

Table 1.2 Consensus DNA sequences

Name	Subclass	Sequence
Chromosome structure		
Scaffold associated region (SAR)[a]	T-box	5'TT(A/T)T(T/A)TT(T/A)TT3'
	A-box	5'AATAAA(T/C)AAA3'
Topoisomerase II cleavage sequence[a]		5'GTN(A/T)A(T/C)ATTNATNN(G/A)3'
Autonomously relicating sequence (ARS)[a]		5'ATTTAT(A/G)TTTA3'
Centromere (CEN)[a]		(see diagram below)
Telomere (TEL)[b]		5'C_n(A/T$_m$) 3', $n >1$, $m = 1$–4
Transcript processing		
Transcription start[b]		5'CTCATCA3'
Introns[b]	5' Splice site	5'AAG:GUAAGU3'
	3' Splice site	5'U$_{16}$GCAG:GU3'
Poly(A) signal[b]	Near-upstream element	5'AAUAAA3'
Control of gene expression		
Translation start[b]		5'UAAAC̲A̲U̲G̲G̲C̲U3'
Proximal promoter elements[b]	TATA box	5'T(C/G)TATA(T/A)A$_{1-3}$(C/T)A3'
	CAAT/ AGGA box	5'(C/T)A$_{2-5}$(C/T)NGA$_{2-1}$(C/T)(C/T)3'
	Legumin box	5'TCCATAGCCATGCAAGCTGCAGAATGTG3'
	Light-responsive element	Box II: 5'GTGTGGTTAATATG3' Box III: 5'ATCATTTTCACT3'
Distal promoter elements[b]	Abscisic acid-responsive element	5'GTACGTGG3'

Centromere (CEN)[a]:

I	II	III
5'TCAC	76–86 bp	CCGAAA3'

[a] Eukaryote data.
[b] Plant data.

telomere-binding protein that may have a protective function. The telomeres may also contain extensive regions of repetitive sequences that are proximal (closer to the centromere) than the simple telomere DNA repeats. In rye this DNA has been estimated to account for 12–18% of the genome.

1.4 DNA methylation

Cytosine residues in the nuclear chromosomes of plants are highly methylated. About 30% of all the cytosine bases in the wheat genome are methylated compared with only 1–7% in animal genomes and it has been estimated that 82% of the CpG dinucleotides of wheat DNA are methylated. In addition to CpG dinucleotides, and unlike in animals, the cytosine in CpXpG trinucleotides is also methylated in plants. The incorporation of 5-methylcytosine into DNA is a post-replicative modification. In animals there are fewer CpG dinucleotides in the genome than would be expected on the basis of the C+G content and the random choice of nucleotides; however in plants there is little evidence for the suppression of CpG dinucleotides. For example, wheat has a C+G content of 45% and the ratio of observed/expected CpG is 0.77 for bulk DNA, whereas in humans with a C+G content of 40% the ratio is only 0.23. DNA methylation has been implicated in several types of non-Mendelian phenomena in plants (see section 1.6, Epimutation).

In the human genome a normal distribution of CpG dinucleotides does occur in short stretches of about 1 kb, at the 5' end of housekeeping genes. Such CpG islands are unmethylated and are thought to mark transcriptionally active genes. Restriction enzymes such as *Bss*HII (GCGCGC), *Sac*II (CCGCGC) and *Not*I (GCGGCCGC), which only cut unmethylated DNA, often cut the human genome in these CpG islands because of the relative enrichment of G+C bases and the unmethylated CpG dinucleotides. Despite the lack of CpG suppression in the cereal genomes, similar islands of unmethylated CpG dinucleotides have been shown to provide landmarks for genes. DNase I-sensitive regions are under-methylated relative to total DNA in pea, barley and maize.

1.5 Structure of nuclear genes

The general structure of plant nuclear genes follows the pattern of other eukaryotic genes that have been studied. Figure 1.3 shows the structure of a generalized plant gene as: (i) chromosomal DNA, (ii) primary transcript (pre-mRNA) and (iii) messenger RNA (mRNA). In outline the chromosome contains sequences 5' (preceding or upstream) of the transcribed sequence called the promoter. This promoter will contain signals (elements) that are important in controlling the expression of the gene. The promoter is followed by the sequence of the primary transcript. This sequence contains signals important for the processing of the primary transcript to produce mRNA. The sequence of the primary transcript will end at a point that is not well defined and will be followed by 3' sequence, which has not been shown to have a significant role in gene expression.

Box 1.7 Codon usage

The term 'codon usage' describes the selective and non-random use of synonymous codons in protein synthesis. Several investigators have noted species-specific patterns to codon usage. In higher plants, monocotyledonary (monocot) and dicotyledonary (dicot) species differ significantly from each

▶

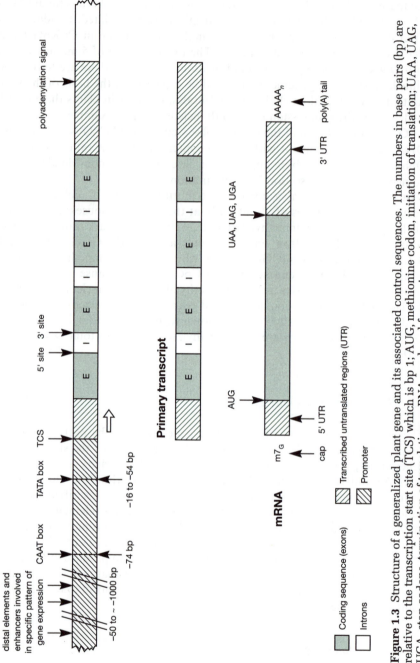

Figure 1.3 Structure of a generalized plant gene and its associated control sequences. The numbers in base pairs (bp) are relative to the transcription start site (TCS) which is bp 1; AUG, methionine codon, initiation of translation; UAA, UAG, UGA, stop codons, termination of translation; mRNA produced from primary transcript by removal of intron sequences between the 5′ and 3′ splice sites.

(Box 1.7 continued)

other. Monocots show a distinct preference for XXC/G codons and have a higher frequency of XCG but a lower frequency of XTA than dicots. It has also been noted that the genes encoding abundant proteins show more codon bias than those encoding less abundant proteins.

Promoter regions

The region of the chromosome lying 5' of the transcribed region is called the promoter region. It contains two types of element. Elements furthest from the transcription start site (distal elements) may be several hundred bases away from the proximal elements, which are normally situated within 75 bp of the transcription start site (TCS). The distal elements have been studied for a number of genes and are responsible for the particular pattern of expression of a gene (see section 1.6). The proximal elements are involved in the general control of transcription, that is they are found in most genes (see section 1.6) and consist of two motifs: the CAAT or AGGA box and the TATA box. Not all plant genes contain a CAAT or AGGA box; where one of these exists, functional studies of mutant sequences suggest that it regulates the level of transcription. The TATA box is closer to the TCS, normally 30 bp 5' of the TCS (sometimes written as –30 bp), and is present in virtually all expressed plant genes. The TATA box is involved in orientation of RNA polymerase II, the enzyme responsible for mRNA synthesis (transcription).

Box 1.8 Enhancers

Enhancers are sequences that can be situated at a considerable distance from the TCS and act in a position- and orientation-independent manner to stimulate the expression of a gene. In the soybean storage protein (conglycinin) gene a sequence of four repeats of A(A/G/C)CCA behaves like an enhancer and can give rise to a 25-fold increase in the level of gene expression.

Following transcription the primary transcript is processed by (i) addition of the modified base, 7-methylguanosine, to cap the 5' end, (ii) removal of introns and (iii) the addition of a poly(A) tail.

Introns

Many plant nuclear genes contain intervening sequences or introns (sometimes called spliceosome introns). These are sequences excised from the primary transcript (pre-mRNA) by a process known as splicing. The basic mechanism of splicing has been established in animal and yeast systems. Splicing proceeds through two steps. In the first step, cleavage occurs at the 5' splice site, with the formation of a covalent attachment of the cut 5' end of the intron to an internal adenosine located within the intron, upstream of the 3' splice site. In the second step, the 3' splice site is cleaved and the two exon (coding) sequences are ligated. Splicing depends on sequence signal recognition by a large complex called the spliceosome. The spliceosome consists of proteins and uracil-rich small nuclear RNAs (UsnRNAs). Although relatively little is known about splicing in plants, sequence similarity of intron sequence signals and UsnRNAs suggests that it occurs by a mechanism similar to that in animals and yeast.

Figure 1.4 shows the size distribution of plant introns in both dicots and monocots. Over 30% of the introns sequenced are between 80 and 100 bp in length whereas in vertebrates only 13% of introns are below 100 bp. This may indicate that plant introns are generally smaller than animal introns.

Box 1.9 Large plant introns

The largest plant introns are intron 1 of *En*-1 mosaic protein gene (4434 nucleotides (nt)) in the monocot *Zea mays* and intron 4 of *Tam*1 gene (4505 nt) in the dicot *Antirrhinum majus*. The efficient splicing of these shows that plants can process large introns.

A comparison of 1334 plant introns produced 5' splice site (AAG:GTAAGT) and 3' splice site (T_{16}GCAG:GT) consensus sequences similar to those found in animal genes. The AG dinucleotide at the 3' end of the intron is conserved in virtually all introns; however about 1% have the 5' dinucleotide GT replaced by

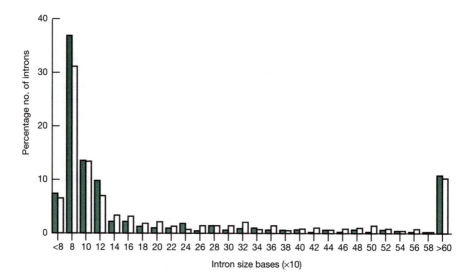

Figure 1.4 Size distribution of plant introns ■ , monocot; □ , dicot. (From Simpson, C, G., Leader, D.J. and Brown, J.W.S. (1993) Characteristics of plant pre-mRNA introns and transposable elements. Reproduced from Croy, R.R.D. (ed.) *Plant molecular biology labfax*, pp. 183–253, with permission from BIOS Scientific Publishers, Oxford.)

GC. Differences between animal and plant introns include (i) the low abundance of C nucleotides upstream of the 3' splice site in plants, (ii) a less prominent polypyrimidine tract and (iii) a high AU content in plant introns. There are differences in the introns of monocot and dicot plants with the dicots having a higher AU content than monocots. It has been shown experimentally that an AU content of 59% is minimal for efficient splicing in dicots. Only 1.6% of dicot introns fall below this figure, whereas 38.2% of monocot introns have less than 59% AU content.

Poly(A) signals

A tract of polyadenylate (poly(A)) is added to the 3' end of most mRNAs. The animal consensus poly(A) signal (AAUAAA), which occurs near the site of polyadenylation, is only found in about 33% of plant genes but a match of four to five of the six bases can be found in another 50%. This means that a significant proportion of plant genes have no AAUAAA-like signal. In addition, when present, most plant transcripts contain more than one polyadenylation signal. Functional studies confirm that plant polyadenylation signals differ from those of mammals and yeast.

Our current understanding of plant poly(A) signals is based largely on studies of four genes (CaMV 19S/35S unit, rbcS-E9 gene, ocs gene, a zein 27 kM$_r$ gene) that come from different sources. Since these studies lead to a similar model, this may be generalized to other plant genes. Plant polyadenylation signals appear to be complex and consist of three elements (Figure 1.5). The most distant from the site of cleavage and polyadenylation of the transcript is called the far-upstream element (FUE). These regions have been identified in functional experiments, by studying their effect on polyadenylation. There is not a great deal of sequence homology; however the motifs UUGUA occur in three of the genes and all FUEs are rich in the UG dinucleotide. The near-upstream elements (NUEs) occur within 40 nucleotides of the polyadenylation site and these may be characterized as AAUAAA-related sequences. There are few functional studies of these sequences but it has been suggested that the variation in this element indicates a novel mechanism of polyadenylation in plants. There is an even less clear definition of the actual cleavage and polyadenylation site and the most important feature is probably the distance from the NUE.

Studies of pea rbcS genes, which have multiple poly(A) sites, indicate that there is no differential site selection under a range of environments and developmental stages. However the results of other workers suggest that in some

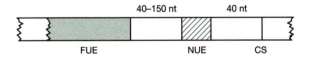

Figure 1.5 Plant gene polyadenylation signals. CS, cleavage/poly(A) site; FUE, far-upstream element; nt, nucleotides; NUE, near-upstream element.

cases there may be preferential use of one polyadenylation site and further experimentation is needed before the significance of the existence of multiple sites can be established.

1.6 Control of gene expression

The processes of protein synthesis in plants follow those of other eukaryotic organisms. The production of a functional plant protein, from the genomic sequence, can be controlled at any of a number of different stages (Figure 1.6). These can be classified as:

1. conformation of chromatin,
2. control of transcription,
3. control of transcript processing and transport
4. mRNA stability,

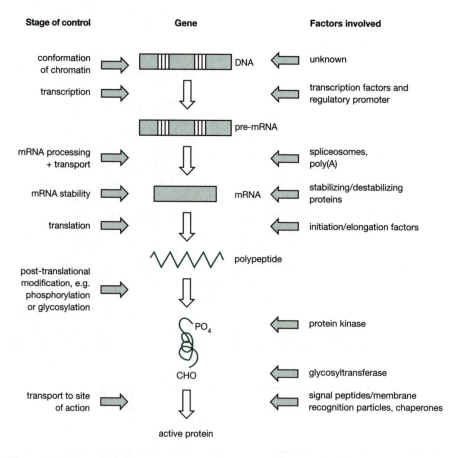

| Stage of control | Gene | Factors involved |

conformation of chromatin — DNA — unknown

transcription — transcription factors and regulatory promoter

pre-mRNA

mRNA processing + transport — spliceosomes, poly(A)

mRNA stability — mRNA — stabilizing/destabilizing proteins

translation — initiation/elongation factors

polypeptide

post-translational modification, e.g. phosphorylation or glycosylation

PO$_4$ — protein kinase

CHO — glycosyltransferase

transport to site of action — signal peptides/membrane recognition particles, chaperones

active protein

Figure 1.6 Control of plant gene expression: stages of control and factors involved.

5. control of translation,
6. control of post-translational processes, such as protein phosphorylation and
7. protein transport.

This section will give an overview of these processes, which will be dealt with in more detail in later chapters (Chapters 9, 10 and 13) where they will be discussed in relation to particular genes.

In common with other eukaryotes, plant cell nuclei contain three forms of RNA polymerase: RNA polymerase I catalyses the synthesis of rRNA; RNA polymerase II catalyses the synthesis of mRNA and is the subject of this section; and RNA polymerase III catalyses the synthesis of transfer RNA (tRNA) and the low molecular mass 5S rRNA.

Chromatin conformation

The first criterion for active transcription of a gene by RNA polymerase II is the availability of the DNA, which in turn is determined by the conformation of the chromatin. Figure 1.7 shows the results of an experiment in which chromatin was extracted from the leaves and the endosperm of barley plants, digested with

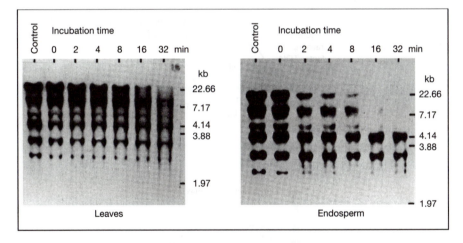

Figure 1.7 Changes in chromatin conformation of genes being transcribed, shown by changes in the sensitivity of the DNA to DNase I digestion. Chromatin samples were extracted from barley leaves given a day/night cycle and developing endosperm tissue and treated with DNase I for various times. The DNA was then cut with *Eco*RI and the fragment size fractionated by electrophoresis. A Southern blot of the DNA fragment was probed with a hordein (storage protein) cDNA probe. The hordein genes are only transcribed in the developing endosperm where the DNA is more degraded by DNase I. (From Steinmüller, K., Batschauer, A. and Apel, K. (1986) Tissue-specific and light-dependent changes of chromatin organisation in barley. *European Journal of Biochemistry*, **158**, 519–525.)

the endonuclease DNase I and subsequently cut with the restriction endonuclease *Eco*RI. The resulting DNA fragments from both sources were separated by
agarose gel electrophoresis, Southern blotted and probed with a storage protein
(hordein) cDNA sequence. This gene is only expressed in the endosperm and
Figure 1.7 shows that the hordein gene DNA (chromatin) extracted from
endosperm is much more sensitive to DNase I digestion than the same sequence
in chromatin taken from leaves. This implies that the chromatin structure of the
hordein gene is different in cells where the gene is being expressed.

Transcription factors

The concept that the transcription of genes is mediated by a multifactor nucleo-
protein transcription complex was established over a decade ago. The current
model, based on yeast, is shown in Figure 1.8. This model predicts that the

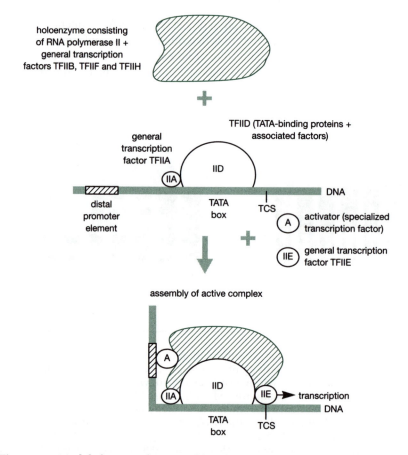

Figure 1.8 Model showing the assembly of an active transcription complex from
RNA polymerase II and both general and specific transcription factors. TCS, transcription start site.

binding and activity of RNA polymerase II requires the cooperative assembly of a number of general transcription factor proteins which bind to the TATA box (TFIIA, TFIIB, TFIIE, TFIIF, TFIIH) with the TATA-binding protein (TFIID). In addition a specific transcription factor (the activator), which is involved in binding to the distal promoter elements, is required for transcription by RNA polymerase II. Together these transcription factors determine a particular pattern of expression for each gene (Figure 1.8).

Box 1.10 Epimutation

The *P* gene in maize encodes a myb-homologous transcription factor that controls the expression of the gene *A1*. This gene, in turn, encodes an enzyme in the biosynthetic pathway of flavonoids and the pigment anthocyanin. Mutations of *P* cause changes in the pigmentation of the maize kernel pericarp, the cob and the floral organs. The allele *P-pr*, which leads to a variable and variegated phenotype in the pericarp, arose from the *P-rr* solid pigmentation allele. The DNA sequence of both the *P-pr* and the *P-rr* allele is identical but *P-pr* has been shown to be more methylated in two regions that flank the transcript-coding region of *P*. *P-pr* shows Mendelian inheritance and is known as an epimutation (that is, it is not a DNA sequence change). The variegated pattern of *P-pr* arises from demethylation of these regions in some cells during development.

Analysis of DNase I sensitivity shows that eight DNase I-sensitive sites are specific for the *P* gene in expressing tissue; however a comparison of *P-rr* and *P-pr* 'expressing' tissues shows that two of these sites are not available in *P-pr* genotypes. Further, these two sites coincide with the methylated sites of *P-pr*. This is evidence that epimutation affects both site-specific DNA methylation and specific local chromatin structure, which is involved in the regulation of *P* expression.

The sequences of distal promoter elements have been identified for a number of genes. Examples include the following.

1. The legumin box (Table 1.2, TCCATAGCCATGCAAGCTGCA-GAATGTC) determines the developmental regulation of the legumin genes during seed development.
2. The light-responsive element, LRE (Table 1.2: Box II, GTGTGGT-TAATATG; Box III, ATCATTTTCACT) is found in a number of genes (ribulose 1,5-bisphosphate carboxylase and the light-harvesting chlorophyll-binding protein) involved in photosynthesis that are switched on by light.
3. The abscisic acid-responsive element, ABRE (Table 1.2, GTACGTGG) is found in the promoter regions of several genes that are regulated by the plant hormone abscisic acid.

Specialized transcription factors that can bind to these distal promoter elements are beginning to be isolated from plants. One such factor, EmBP-1, was cloned from wheat and found to belong to the bZIP protein family of transcription factors. This factor interacts specifically with the ABRE (Table 1.2). A summary of plant transcription factors and their promoter binding sites is shown in Table 1.3.

Table 1.3 Plant transcription factors with known function

Protein	Species	Motif	Regulatory function	DNA-binding site
Isolated by mutant analysis				
C1	Maize	Myb-like		?
P	Maize	Myb-like	Anthocyanin biosynthesis	?
Rc	Maize	HLH		?
Bc	Maize	HLH		?
O2	Maize	bZIP	22-kM$_r$ zein biosynthesis	GATGACGTGA
Kn1	Maize	Homeodomain	Leaf development	?
DEFA	*Antirrhinum*	SRF-like	Flower development	?
AG	*Arabidopsis*	SRF-like	Flower development	?
Gl1	*Arabidopsis*	Myb-like	Trichome development	?
Isolated from cDNA libraries				
TGA1a	Tobacco	bZIP	as-1-dependent expression	TGACGTA
TGA1b	Tobacco	bZIP	Histone genes	TGACGT
TAF-1	Tobacco	bZIP	G-box-dependent expression	CCACGTGG
CPRF-1[a]	Parsley	bZIP	Light regulation	TTCCACGTGGCCA
HBP-1a[a]	Wheat	bZIP	Histone genes	ACGTA
EmBP-1	Wheat	bZIP	ABA induction	GTACGTGG
GT-1	Tobacco	?	Light regulation	GTGTGGTTAATATG
GT-2	Rice	?	Light regulation	GCGGTAATT
EPF-1	Petunia	Zinc finger	Flower development	TGACAGTGTCA

[a] Part of gene family.

bZIP, basic leucine zipper; HLH, helix–loop–helix structure; Myb-like, homology to Myb animal transcription factor; SRF, serum response factor.

Studies of the control of gene expression in plants at post-transcriptional levels are still in their infancy but there is a growing number of studies in which they are implicated. For example, a number of cereal genes, whose mRNA levels increase at low growth temperature, were shown not to be up-regulated by increases in the rate of transcription. The importance of these post-transcriptional controls is becoming widely recognized as more examples appear in the scientific literature.

Major Learning Objectives for Chapter 1

1. Knowledge of the range of variation in plant nuclear genome size and its relationship with repetitive DNA sequences.
2. Knowledge of chromatin structure and its role in the control of gene expression.
3. Knowledge of DNA methylation and its role in the control of gene expression.
4. Outline of the basic steps in gene expression and the points at which control may occur.
5. Knowledge of basic plant gene structure and the sequence of motifs important in the control of transcription and transcript processing.
6. Understand the interaction of promoter sequence elements and transcription factors in gene expression.

Further reading

This chapter covers a broad range of topics and good reviews of this material are limited.

MEYEROWITZ, E.M. (1994). Structure and organisation of the *Arabidopsis thaliana* nuclear genome. In Meyerowitz, E.M. and Sommerville, C.R. (eds) *Arabidopsis*, pp. 21–36. Cold Spring Harbor Laboratory Press, New York.
This reference describes the genome of the important model plant used for many molecular genetic studies that feature in this text.

MOORE, G., ABBO, S., CHEUNG, W., FOOTE, T., GALE, M., KOEBNER, R., LEITCH, A., LEITCH, I., MONEY, T., STANCOMBE, P., YANO, M. and FLAVELL, R. (1993). Key features of cereal genome organisation as revealed by the use of cytosine methylation-sensitive restriction endonucleases. *Genomics*, **15**, 472–482.
A good reference for methylation.

SLATER, R.E and GRAY, J.C. (1991). Chromatin structure of plant genes. *Oxford Surveys of Plant Molecular and Cell Biology*, **7**, 115–142.
A useful review of chromatin structure and the review closest to the subject matter of this chapter.

The following references contain compilations of sequence data and are a useful source of this information.

ARUMUGANATHAN, K. and EARLE, E.D. (1991). Nuclear DNA content of some important plant species. *Plant Molecular Biology Reporter*, **9**, 208–218.

HUNT, A.G. (1994). Messenger RNA 3' end formation in plants. *Annual Review of Plant Physiology and Plant Molecular Biology*, **45**, 47–60.

JOSHI, CP. (1987). An inspection of the domain between the putative TATA box and the translation start site in 79 plant genes. *Nucleic Acids Research*, **15**, 6643–6653.

KATAGIRI, F. AND CHUA, M-H. (1992). Plant transcription factors: present knowledge and future challenges. *Trends in Genetics*, **8**, 22–27.

SIMPSON, C.G., LEADER, D.J. and BROWN, J.W.S. (1993). Characteristics of plant pre-mRNA introns and transposable elements. In Croy, R.R.D (ed.) *Plant molecular biology labfax*, pp. 183–253. Bios Scientific Publishers, Oxford.

Chapter 2

The inheritance of nuclear genes

2.1 Inheritance of nuclear genes

The rules of inheritance of the nuclear genes of plants are the same as those that apply to the inheritance of nuclear genes in other eukaryotic organisms; however the details of fertilization and embryogenesis in higher plants differ significantly from other organisms.

Higher or flowering plants belong to the class Angiospermae and in these plants there is an alternation of two generations in the life cycle. The flower-bearing generation or sporophyte, with which we are familiar, arises from the development of a zygotic cell produced by the union of a male and a female gamete. In diploid flowering plants each sporophyte cell contains two copies of the nuclear genome and hence two copies of each gene. One copy of each gene comes from the male (pollen-producing) parent and the other from the female (ovule-producing) parent.

The alternate generation (gametophyte) is haploid and is very reduced. The gametophyte generation arises from mitotic division of a haploid cell produced by meiosis. The male gametophyte consists of the few cells produced in the germinating pollen grain, one of these cells being the male gamete. The female gametophyte consists of the embryo sac, which resides in the ovule and is normally composed of about seven cells (Figure 2.1). One of these cells is the female gamete or egg. More information about pollination and fertilization is given in Chapter 12.

Box 2.1 Endosperm

One feature of higher plant reproduction is the production of an endosperm, which serves as a source of nutrients for the developing embryo and is of considerable commercial importance since it is the most important component of yield in many seed crop plants (cereals). In outline, the sporophytic embryo is formed by the union of the female egg cell of the embryo sac and a male gamete from the pollen tube. The endosperm develops from another, large, cell in the embryo sac that contains two haploid (polar) nuclei. These two nuclei coalesce and combine with a further haploid male nucleus from

▶

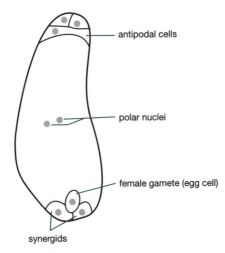

antipodal cells

polar nuclei

female gamete (egg cell)

synergids

Figure 2.1 Female haploid gametophyte or embryo sac. There are seven haploid nuclei in the embryo sac: three are in the antipodal cells, whose function is not known, one nucleus is in the female gamete and two are in the associated synergid cells, which are thought to aid fertilization. The remaining two nuclei (the polar nuclei) fuse to form a diploid nucleus, which combines with the second male nucleus to form a triploid nucleus that divides to give rise to the endosperm.

(Box 2.1 continued)

the pollen tube to produce a triploid cell. This cell divides by triploid mitosis to form endosperm tissue. This means that endosperm cells contain two copies of each gene from the female parent and one copy of each gene from the male parent.

Alternative forms of a gene arise by mutation and are known as alleles. Many mutations lead to a loss of function of the gene and for genes that affect the phenotype (or form) of the sporophyte (flower-bearing plant) these mutations are normally recessive. This means that plants containing one normal (wild-type) allele and one mutant recessive allele of the gene have a normal phenotype. Such plants are said to be heterozygous. Plants that contain identical alleles of a gene are said to be homozygous.

Box 2.2 Genotype, genome and karyotype

The term 'genotype' describes the genetic constitution of a plant. It is not synonymous to the genome in two respects:

1. The genotype describes the particular alleles of each gene present in the plant.
2. The genome usually refers to all the DNA present in the haploid set of chromosomes, including repetitive and structural DNA.

▶

(Box 2.2 continued)

The other useful term is 'karyotype', which describes the shape and number of the haploid set of chromosomes as seen with a light microscope.

Figure 2.2 shows the pattern of inheritance of the alleles of two genes, where these genes reside on different chromosomes. Meiotic (reduction) division produces haploid cells and the segregation of chromosomes during meiosis ensures that each haploid nucleus contains one copy of each chromosome in the nuclear genome. Alleles of genes that are present on different chromosomes will segregate independently during meiosis. This means that during meiosis in a heterozygous plant both the parental combination of alleles (*AB* and *ab*) and the non-parental (*Ab* and *aB*) combination of alleles will occur with equal frequency in the products of meiosis. If such a heterozygous plant is crossed with a homozygous recessive plant the phenotype of the resulting test-cross progeny will be determined by the gametes from the heterozygous parent. The four possible phenotypes will therefore occur with equal frequency in this test-cross progeny.

If two genes are present on the same chromosome they tend to be tied together during the segregation of chromosomes during meiosis. The alleles of

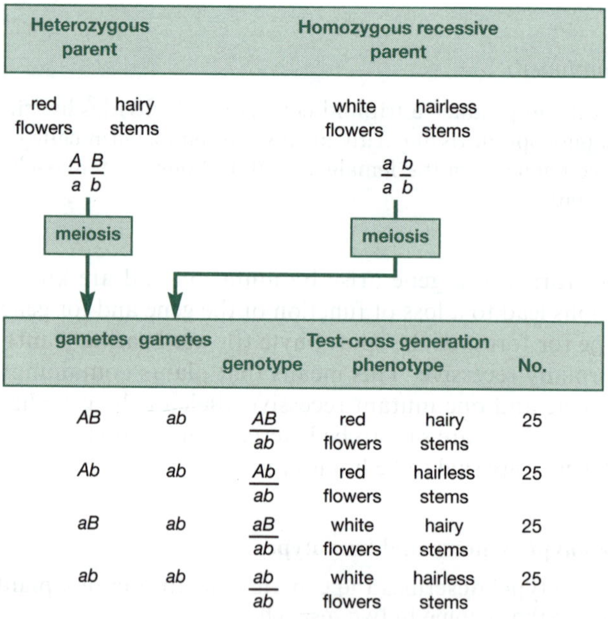

Figure 2.2 A test cross to show the pattern of inheritance of the alleles of two genes situated on different chromosomes. Alleles determining flower colour: *A*, dominant, red phenotype; *a*, recessive, white phenotype. Alleles determining stem hairs: *B*, dominant, hairy phenotype; *b*, recessive, hairless phenotype. The heterozygous parent is produced by crossing a homozygous red flower, hairy stem (*A/A*, *B/B*) plant with a homozygous white flower, hairless stem (*a/a*, *b/b*) plant.

two such genes will only recombine into a novel non-parental combination if chromatid breakage and exchange occurs between the genes during the prophase of meiosis, when homologous chromosomes are paired. The frequency of these breakage and exchange events is a measure of the distance between the genes and is calculated as the percentage of recombinant (non-parental) gametes produced by heterozygotes. The production of 1% recombinant gametes defines 1 map unit (1 map unit is sometimes referred to as 1 centimorgan (cM)).

Figure 2.3 shows the pattern of inheritance of alleles of two genes, where these genes are 10 map units (10 cM) apart on the same chromosome. In this cross the four test-cross progeny phenotypes are not produced in equal frequency. The number of progeny inheriting a gamete from the heterozygote that contains the parental combination of alleles is greater (90%) than those progeny inheriting non-parental (recombinant) gametes (10%).

Box 2.3 Linkage group

Genes situated on the same chromosome are said to be linked and form a linkage group.

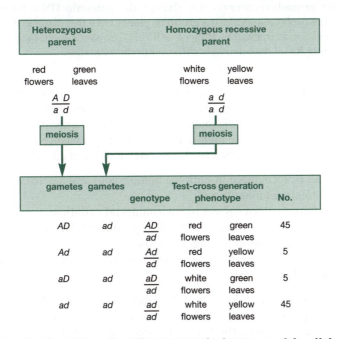

Figure 2.3 A test cross to show the pattern of inheritance of the alleles of two genes linked on the same chromosome. Alleles determining flower colour: *A*, dominant, red phenotype; *a*, recessive, white phenotype. Alleles determining leaf colour: *D*, dominant, green phenotype; *d*, recessive, yellow phenotype. The heterozygous parent is produced by crossing a homozygous red flower, green leaf (*AD/AD*) plant with a homozygous white flower, yellow leaf (*ad/ad*) plant.

2.2 Restriction fragment length polymorphism (RFLP) linkage maps

One area where molecular genetics is used as an aid to conventional plant breeding is in the use of RFLPs.

Box 2.4 RFLP: method of detection

RFLP can be determined by using the ability of cloned DNA sequences to cross-hybridize with homologous DNA sequences that have been immobilized on a membrane.

   If genomic DNA is extracted from a plant and cut with a restriction endonuclease, a range of different-sized fragments will be produced depending on the distances between the restriction endonuclease recognition sites within the genome. The resulting DNA fragments are size fractionated by gel electrophoresis, denatured (to make them single stranded and therefore available for hybridization) and Southern blotted on to a membrane. The cloned DNA sequence is labelled, often by incorporating a ^{32}P-labelled base into the sequence *in vitro*, and this DNA is known as the probe. Denatured radiolabelled probe DNA is allowed to hybridize with DNA on the membrane and, after appropriate washing, the region of hybridization can be visualized by either exposing an X-ray film or a phosphoimager to the membrane. This procedure reveals the size of the genomic DNA fragment containing the probe sequence. If a group of plants have this sequence in different-sized genomic fragments, they have RFLP. The same membrane can be reused to study RFLP detected by a different probe sequence.

Point mutations in the restriction endonuclease recognition sequence and deletions, insertions or inversions may all cause variation in the length of the restriction fragment. Like any nuclear genetic mutation the restriction fragment is inherited as a Mendelian marker, but unlike mutations recognized by a change in the phenotype RFLP is identified at the DNA level. This has a number of advantages:

1   RFLPs can be recognized in both the homozygous and heterozygous state (co-dominant inheritance);
2   RFLP patterns are not subject to environmental variation;
3   unlike many gene mutations identified by phenotype, RFLPs are identified in all tissues at all stages of growth;
4   only small amounts of tissue are needed and the technique can be standardized between laboratories.

The probe DNA may be a known cloned gene (either cDNA or genomic sequence) or the probe may consist of anonymous genomic DNA with perhaps no function. There are over 100 restriction endonuclease enzymes available and this wide choice of cutting sites, together with no limitations on the DNA to be used as probes, means that comprehensive RFLP maps covering all the genome have been produced for a number of plant species, including *Arabidopsis*, tomato and maize. In *Arabidopsis* the RFLP markers are, on average, 1.5 cM (300 kb) apart.

Figure 2.4 illustrates the different patterns of inheritance of RFLP markers seen with two contrasting situations: independent assortment (fragments on separate chromosomes) and complete linkage of the DNA fragments. Two probe sequences are used (A and B) which hybridize to short internal sequences in the restriction fragments. Two parent plants are shown: parent P1 is heterozygous for restriction fragments hybridizing to each probe, whereas parent P2 is homozygous for the long probe B fragment and for the short probe A fragment. These parents are crossed and the progeny scored for the

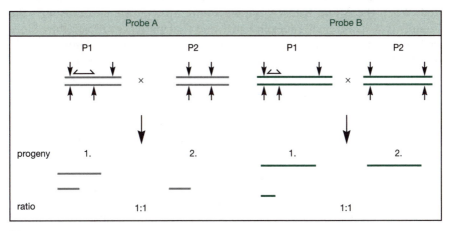

(a)

(b)

Figure 2.4 Patterns of inheritance of restriction fragment length polymorphism (RFLP) for two loci that are either situated on separate chromosomes and segregating independently or completely linked assuming that both long fragments are as the same chromosome in parent 1. (a) Use of a single probe; (b) use of both probes. ⤳ :region hybridizing with probe; ↓ restriction endonuclease site; P1, parent 1; P2, parent 2.

size of the restriction fragments hybridizing to each probe. If only one probe is used to analyse the plants, it can be seen that for probe A progeny will inherit either the long or the short fragment from parent P1 and a short fragment from homozygous parent P2, giving heterozygous 'long/short' plants and homozygous 'short' plants in a 1:1 ratio (Figure 2.4a). A similar pattern of inheritance can be seen for fragments hybridizing to probe B.

If each progeny plant is scored for the RFLP pattern of both probes, the total number of fragments measured will vary from four to two depending on the combination of A fragments and B fragments (Figure 2.4b) that are inherited. When the A and B DNA sequences are on separate chromosomes the fragments will segregate independently and four patterns are possible. Since each pattern has an equal chance of occurring among progeny plants there will be a 1:1:1:1 ratio of these types of plants. Figure 2.4b also shows the pattern of inheritance seen when the two DNA sequences are completely linked, assuming that in the heterozygous parent P1 the long A fragment is linked to the long B fragment and the short A and B fragments reside on the other chromosome. In this case every gamete produced by parent P1 containing the long A fragment will also contain the long B fragment, and similarly the two short fragments, A and B, will always be inherited together. Therefore there are only two types of progeny from the cross, each with three fragments, and these types will occur in a 1:1 ratio.

RFLP markers, like those illustrated in Figure 2.4, can also be mapped relative to phenotypic markers. Figure 2.5 shows the pattern of segregation of a

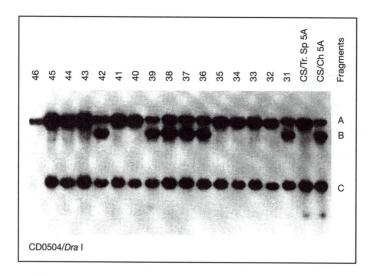

Figure 2.5 Autoradiograph of *Dra* I cut wheat DNA from two parental lines, CS/Ch5A and CS/Tr.Sp.5A and a sample of segregating progeny, probed with the RFLP marker CD0504. (From Galiba G., Quarrie, S.A., Sutka, J., Morgounov, A and Snape, J.W. (1995) RFLP mapping of the vernalisation (*Vrn*1) and frost resistance (*Fr*1) genes on chromosome 5A of wheat. *Theoretical and Applied Genetics*, **90**, 1174–1179.)

Dra I fragment (B) identified in a cross between two wheat parent lines by probing with the anonymous genomic DNA clone (CD0504), which co-segregates with a phenotypically identified genetic difference. Parent CS/Ch5A is heterozygous for the presence of a *Dra* I restriction site, which gives rise to the *Dra* I fragment, not present in the homozygous parent CS/Tr.Sp.5A. Of the 16 progeny tested six plants inherited the *Dra* I site from CS/Ch5A (in a larger sample this would be a 1:1 ratio). This *Dra* I site (fragment B) co-segregates with alleles of the gene, *Vrn1*, which determines vernalization requirement. Vernalization (a low temperature treatment) is necessary for flower development in winter cereals and many other over-wintering plants (see section 11.2).

Box 2.5 Segregation in hexaploid wheat

Wheat is a hexaploid, having three diploid sets of chromosomes. DNA probes commonly reveal a number of bands in wheat genomic Southern blots; this is due to hybridization with the equivalent gene or sequence in all six of the homologous chromosomes. However, because pairing between chromosomes is regular in each diploid set, individual RFLP fragments can show regular diploid Mendelian inheritance.

2.3 RFLP maps and mapping quantitative trait loci (QTL)

Many characters (traits) that are important in crop improvement, such as frost tolerance, exhibit continuous variation. It has been established that the quantitative pattern of inheritance of these traits arises from the segregation of the alleles of multiple genes which are often modified by environmental factors. The systematic mapping of genes contributing to a continuously variable trait (QTL) was not feasible before RFLP markers, because the inheritance of an entire genome could not be studied with phenotypic genetic markers. The existence in a number of plant species of RFLP linkage maps covering the entire genome and the adoption of mathematical techniques used in human linkage analysis has enabled some QTL to be mapped in plants.

The basic methodology for mapping QTL is to cross two inbred lines (homozygous genotypes) that differ substantially in a quantitative trait. Segregating progeny, normally produced by a back-cross (F_1 × parent), are scored both for the quantitative trait and for a number of RFLP markers. At each position on the genome defined by the RFLP markers, the most likely phenotypic effect of the putative QTL is computed together with the odds ratio. This ratio is the chance that the data would arise from QTL with this effect divided by the chance that it would arise given no linked QTL. The $\log_{10}$ of this odds ratio is known as the lod score and it measures the strength of evidence in favour of the existence of QTL at this position.

QTL for the soluble-solids concentration of tomato fruits have been mapped on chromosome 6 of tomato. Figure 2.6 shows the QTL likelihood map indicating the lod score for different positions along the chromosome. The height of the curve indicates the strength of the evidence for the presence of QTL at each position and the horizontal line gives a threshold that the lod score must exceed to allow QTL to be identified.

A mathematical treatment of QTL mapping is outside the scope of this text but can be found in Lander and Botstein (1989). This technique is significant in the context of plant molecular genetics because once QTL are mapped, RFLP markers allow the rapid construction of near-isogenic lines. Flanking markers can be used to retain the QTL and the remaining markers used to select individuals that have the same genotype as a recurrent parent. This will produce individuals whose genotype only differs at the single QTL. Using these isogenic lines the techniques of molecular genetics can then be used to study the QTL.

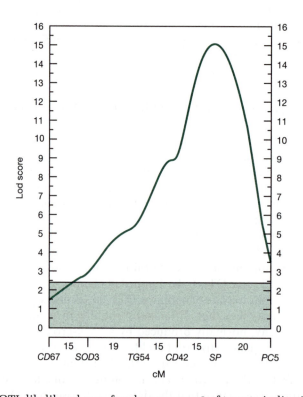

Figure 2.6 QTL likelihood map for chromosome 6 of tomato indicating lod scores for soluble-solids concentration. The distance between the RFLP marker loci *CD67*, *SOD3*, *TG54*, *CD42*, *SP* and *PC5* are given in cM. The lod score threshold of 2.4 is marked with a horizontal line. (From Patersen, A.H., Lander, E.S., Hewitt, J.D., Petersen, S., Lincoln, S.E and Tanksley, S.D. (1988) Resolution of quantitive traits into Mendelian factors by using a complete linkage map of restriction fragment length polymorphism. *Nature*, **335**, 721–726.)

2.4 Karyotype evolution

Comparative genome mapping has progressed rapidly in recent years, partly as a result of the use of molecular markers in genetic analysis. These developments have led to greater understanding of the evolutionary relationship of certain groups of plants. The cereals are one such group where the genetic and physical maps of chromosomes can be compared. Cereals such as barley, wheat, rye, rice and maize are all members of the family Gramineae (grasses). They have different numbers of chromosomes and, in some cases, very different sizes of genome; nevertheless they display a surprising conservation of gene composition and gene order. The co-linearity of these genomes exists both for single-copy genes as well as for the arrays of tandemly repeated genes (e.g. 18S–28S and 5S rRNA genes) and is only broken down where translocations have occurred during evolution.

Figure 2.7 shows the rye karyotype and its relationship to the chromosomes of the D genome of wheat. This figure shows that a relatively small number of translocations may have occurred during the evolution of these two species.

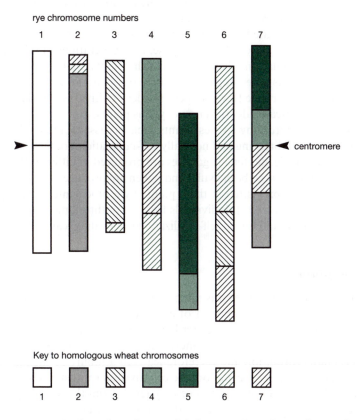

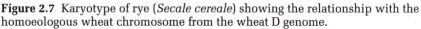

Figure 2.7 Karyotype of rye (*Secale cereale*) showing the relationship with the homoeologous wheat chromosome from the wheat D genome.

Thus rye and wheat chromosome 1 are identical; however rye chromosome 2 is not identical to the wheat chromosome 2 and the upper (short) chromosome arm contains sequences homologous to wheat chromosomes 6 and 7. These data have important implications for the progress of molecular genetic analysis in related species because it suggests that the genomic information from one species may be used to interpret other genomes.

2.5 Genetic manipulation of plant genomes

There are a number of features of the biology of flowering plants that are important in relation to their genetic manipulation. These will be considered in more detail in later chapters but are presented in outline here.

Breeding systems

There are a variety of breeding systems found among angiosperm species. Some plant species are inbreeding, that is the plants normally self-fertilize. One important consequence of inbreeding is the homozygosity of resulting generations. It can be calculated that from a population where 100% of the plants are heterozygous for the alleles of a single gene, four generations of self-fertilization will produce a population where 93.75% of the plants are homozygous. Barley (*Hordeum vulgare*) is an example of this type of plant and pollination has normally occurred before the flowers open. Modern cultivars of barley are largely composed of genetically identical homozygous plants.

In contrast to inbreeding species, plants such as cassava, maize, *Brassica* spp. and tobacco are outbreeding and normally self-pollination does not occur. In cassava and maize, male and female gametes are produced in different flowers but in *Brassica* spp. and tobacco both ovules and pollen are produced in the same flower. Self-fertilization is prevented in these plants by a biochemical incompatibility mechanism (see Chapter 12). Individuals from an outbreeding species are usually highly heterozygous and forced self-fertilization (inbreeding) causes loss of vigour.

Haploid plants

In a number of plant species it is possible to manipulate the developing pollen grains to induce the development of a sporophyte plant, which is therefore haploid. Such haploid plants are sterile but the number of chromosomes may be doubled by treatment of the meristems with colchicine. This chemical inhibits spindle formation and therefore leads to a failure of the regular segregation of chromosomes during somatic cell division (mitosis). These plants are known as doubled haploids and since they result from the doubling of a single haploid set of chromosomes they are 100% homozygous.

If the 'parent' plant used as a source of developing pollen is heterozygous each doubled haploid plant produced will be genetically different, since these plants represent the haploid products of different meioses where recombination of alleles of different genes may have occurred. Analysis of the inheritance of the alleles of genes using doubled haploid plants produced in this way is equivalent to the direct analysis of the gametes produced by a heterozygote; this can be powerful tool of analysis in some inbreeding species.

Recombinant inbreds for RFLP mapping

Recombinant inbred lines are produced by making a cross between two homozygous but genetically different plants to produce a heterozygous F_1 generation. The F_1 plants are self-fertilized to produce an F_2 generation where each individual will be genetically distinct. Individual F_2 plants are self-fertilized, as are the resulting successive generations until homozygosity of individual plants is achieved. At this stage the genotype of each recombinant inbred line is fixed. Mapping in this population of recombinant inbred lines is simple. Each line is typed for the parental restriction fragment marker (allele) it received for every locus to be mapped. When two loci are linked, fragments (alleles) from the same parent will occur together, in the population of lines, more frequently than if they were segregating independently.

Recombinant inbred populations have two advantages over analysing back-cross progeny. The first advantage comes from the fact that each homozygous recombinant inbred line can be propagated indefinitely. With back-cross progeny each plant is unique and once the DNA has been used a new segregating progeny must be characterized for all the loci being mapped. This is a significant drawback in big mapping programmes. The second advantage of recombinant inbred lines is that several meioses (and hence several rounds of recombination) will have occurred during the production of the several selfing generations needed to give homozygosity. This means that there is a greater chance of recombination occurring between linked loci and this is an advantage in studying loci which are close together.

Cytogenetics

Polyploid species are more common in plants than in animals, and many plants have evolved mechanisms that allow them to tolerate extra chromosomes. This phenomenon has led to the development of a number of cytogenetic stocks for the analysis of cereal genomes.

One example of this type of analysis is shown in Figure 2.8. A series of wheat lines have been constructed which, in addition to the normal hexaploid complement of 42 wheat chromosomes, contain one arm of a single barley chromo- some. The barley chromosome arms are in the form of ditelosomic chromosomes, in which a single chromosome arm is duplicated either side of the original centromere. Barley has a nuclear genome of seven chromosomes but there are only 13 barley ditelosomic addition lines in wheat because plants containing the long arm of chromosome 1 are sterile. DNA has been extracted from each of the addition lines as well as from barley and the parent wheat cultivar. This DNA was digested with the restriction enzyme *Eco*RI, and the resulting fragments size fractionated by agarose gel electrophoresis and Southern blotted onto a nylon membrane. The membrane was subsequently probed with a barley gene whose chromosome linkage group was sought.

Figure 2.8 shows that cognates of this barley gene exist in the wheat genome since wheat DNA fragments will hybridize to the barley gene probe. Figure 2.8 also shows that barley chromosomal fragments containing the gene can be identified because they differ in size from the wheat fragments. These two fragments (major = 7 kb, minor = 13 kb) are only present in barley and in the wheat ditelosomic addition line 4HL, revealing that this gene occurs in the long arm of barley chromosome 4. This single Southern blot membrane can be reused to identify the chromosome arm location of several barley genes.

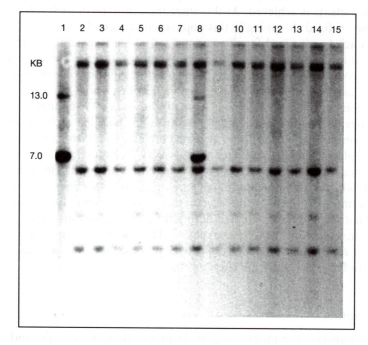

Figure 2.8 Southern blot analysis of barley ditelosomic addition lines of wheat comapred with the barley parent (cv. Betzes, lane 1) and the background wheat parent (cv. Chinese Spring, lane 15) showing that the barley gene *blt*101 lies on the long arm of chromosome 4. Lanes 2–14, barley ditelosomic addition lines: (2) 1HS (chromosome 1H, short arm); (3) 2HS; (4) 2HL (chromosome 2H, long arm); (5) 3HS; (6) 3HL; (7) 4HS; (8) 4HL; (9) 5HS; (10) 5HL; (11) 6HS; (12) 6HL; (13) 7Hα; (14) 7Hβ. Barley chromosome nomenclature in accordance with homoeologous wheat chromosomes: 1H, barley chromosome 5; 2H, barley chromosome 2; 3H, barley chromosome 3; 4H, barley chromosome 4; 5H, barley chromosome 7; 6H, barley chromosome 6; 7H, barley chromosome 1. (After Goddard, N.J., Dunn, M.A., Zhang, L., White, A.J., Jack, P.L. and Hughes, M.A. (1993) Molecular analysis and spatial expression of a low-temperature-specific barley gene, blt101. *Plant Molecular Biology*, **23**, 871–879.)

Major Learning Objectives for Chapter 2

1. Understand the terms phenotype, genotype, karyotype, homozygous, heterozygous.
2. Revise independent assortment and linkage.
3. Understand the important uses of restriction fragment length polymorphism (RFLP).
4. Be acquainted with examples of experimental and evolutionary changes in plant nuclear genomes and karyotypes.

Further reading

BURR, B. and BURR, F.A. (1991). Recombinant inbreds for molecular mapping in maize. *Trends in Genetics*, **7**, 55–60.
    A good description of the production and method of using recombinant inbreds for molecular mapping; also contains an RFLP map of maize produced using the recombinant inbred lines.
LANDER, E.S. and BOTSTEIN, D. (1989). Mapping Mendelian factors underlying quantitative traits using RFLP linkage maps. *Genetics*, **121**, 185–199.
    An explanation of the calculation and use of lod scores for mapping QTL.
MOORE, G., DEVOS, K.M., WANG, Z. and GALE, M.D. (1995). Grasses, line up and form a circle. *Current Biology*, **5**, 737–739.
    A review of the recent developments in cereal genome analysis and the prospective use of a composite map of the ancestral grass genome.
TANKSLEY, S.D. (1993). Development and applications of a molecular linkage map in tomato. In Yoder, J. (ed.) *Molecular biology of tomato*, pp. 19–35. Technomic Publishing Co., Lancaster, PA.
    A short presentation of the status of the RFLP linkage maps of tomato, together with an outline of the applications of the molecular map in molecular studies of tomato rather than in tomato breeding.

Chapter 3

Chloroplast genome

3.1 Chloroplast structure and function

Chloroplasts are organelles that differentiate in the light from cytoplasmic organelles called proplastids present in meristematic cells. Chloroplasts are 5–10 μm long, lens-shaped structures surrounded by a double membrane (Figure 3.1) and leaf mesophyll cells can contain up to 100 chloroplasts.

The chloroplast double membrane surrounds the stroma, which contains the enzymes for carbon dioxide fixation, amino acid biosynthesis and lipid metabolism. Within the stroma is the thylakoid membrane system, which contains the proteins for the light reactions of photosynthesis. The biochemistry of photosynthesis will not be covered here but a reference dealing with this topic is given at the end of the chapter.

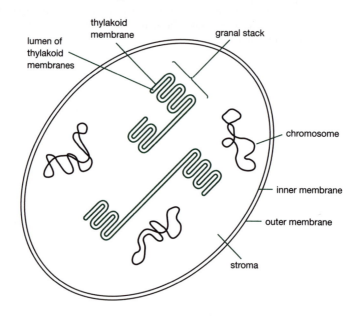

Figure 3.1 Chloroplast structure

Chloroplasts are replicative organelles and contain a number of copies of a double-stranded circular DNA chromosome. The number of copies of this chromosome in each chloroplast varies between cells. In the mesophyll cells of young leaves there may be up to 100 whereas the chloroplasts of old leaves have 20–30. The number of copies of the chloroplast chromosome in individual leaf cells can therefore vary between 500 and 10 000 and may make up 10–20% of the DNA present in leaf cells.

Chloroplast chromosomes lie within the stroma and a number of features of their structure resemble prokaryotic chromosomes. They are circular DNA molecules which, unlike nuclear chromosomes, are not complexed with histones. Replication of the chloroplast genome and its distribution between daughter proplastids is a complex and ill-defined process. In most plants the DNA molecule is between 120 and 160 kb long. The entire chloroplast chromosome of tobacco, rice and the liverwort, *Marchantia polymorpha*, has been sequenced.

Chloroplast genomes have been classified into three types. Two groups of land plants, namely the gymnosperms, Pinaceae, and a group of legumes (including peas and broad bean) have chloroplast chromosomes without an inverted repeat (Group I). Most land plants, including all other angiosperms, have chloroplast genomes containing a large (6–76 kb) inverted repeat (IR); these are Group II genomes. The alga *Euglena* has three tandem repeats in its Group III chloroplast genome.

> **Box 3.1 Evolution of legume chloroplast genomes**
>
> The phylogeny of the unusual chloroplast genomes of some legume species has received a lot of attention. The available evidence suggests that these species form a monophyletic group, that is, they appear to be derived from a single common ancestor which lost one copy of the inverted repeat.

3.2 Chloroplast genome organization

Chloroplast genomes contain between 120 and 140 genes (Figure 3.2). Our knowledge of these genes comes from genetic analysis, studies of proteins synthesized by isolated chloroplasts, RNA–DNA hybridization (which identified some rRNA and tRNA genes) and cloning together with sequence analysis. Figure 3.2 shows the genetic map of the tobacco chloroplast chromosome, which is representative of the gene order found in land plant chloroplasts.

Protein synthesis

The genetic information for much of the apparatus involved in the synthesis of chloroplast-encoded proteins is present in the chloroplast genome, with four genes for rRNA present in the large IR. The other chloroplast protein synthesis genes identified include genes for ribosomal proteins, 30 tRNA genes and

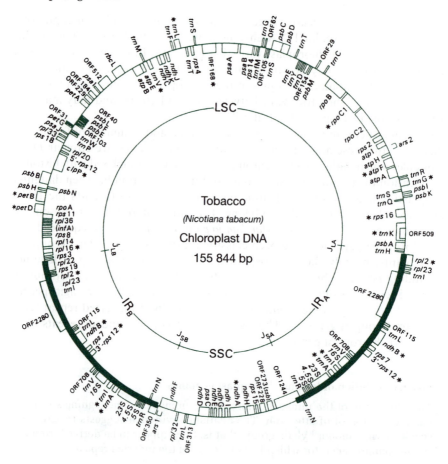

Figure 3.2   Gene map of the tobacco chloroplast genome. Genes shown inside the circle are transcribed clockwise, genes on the outside are transcribed anticlockwise. Asterisks denote split genes. The major open reading frames are included. IRF, intron-containing reading frame; IR, inverted repeat; LSC, large single-copy region; SSC, small single-copy region; J, junctions between IR and LSC and SSC. (From Sugiura, M. (1992) The chloroplast genome. *Plant Molecular Biology*, **19**, 149–168.)

an RNA polymerase gene together with protein synthesis coupling and elongation factors. It is considered that with 'wobble', the 30 tRNA genes are enough for the complete translation of chloroplast mRNA.

Chloroplast protein synthesis resembles the prokaryote mechanism. Thus the chloroplast ribosomes (70S) are smaller than eukaryote cytoplasmic ribosomes (80S) and unlike the cytoplasmic ribosomes are inhibited by the antibiotic chloramphenicol. Table 3.1 shows the rRNAs of maize chloroplast ribosomes. The 23S and 16S molecules are similar to the equivalent molecules from prokaryote ribosomes. The 5S molecule is present in all ribosomes except those of mitochondria in fungi and animals but the 4.5S rRNA molecule is unique to

Table 3.1 Chloroplast ribosomal RNA molecules of maize

	rRNA molecule	No. of nucleotides	Remarks
Large subunit	23S	2860	Similar to prokaryote ribosomes
	5S	122	In all ribosomes except mitochondria of fungi and animals
	4.5S	95	Unique to chloroplasts of higher plants
Small subunit	16S	1450	Similar to prokaryote ribosomes

chloroplasts. Chloroplast ribosomes contain about 60 different proteins and about one-third of these are thought to be encoded by chloroplast DNA. Figure 3.3 shows a comparison of the genes in the tobacco chloroplast *rpl*23 operon, which encodes chloroplast ribosomal proteins, with the homologous genes in the *Escherichia coli* S10, *spc* and α operons. This demonstrates a similar arrangement of these genes and suggests that the genes in *E. coli* and tobacco may have evolved from a common ancestor.

Introns

The *rpl*2 and *rpl*16 tobacco chloroplast ribosomal genes contain an intron, as do several other chloroplast genes. Table 3.2 shows a list of higher plant chloroplast genes that have been found to contain introns. Most of these genes contain a single intron. This table also demonstrates that not all chloroplast genomes have the same structural organization; thus the presence of introns in *rpl*2, *rpo*C1 and *clp*P is not universal in higher plants. The introns found in higher plant chloroplasts are self-splicing and classified into two groups (see also section 4.2 for mitochondrial genome introns). Group I introns form a secondary structure found in fungal mitochondrial gene introns; the *trn*L and 23S rDNA chloroplast introns fall into this class. The other chloroplast genes have Group II introns, which are also designated by their secondary structure.

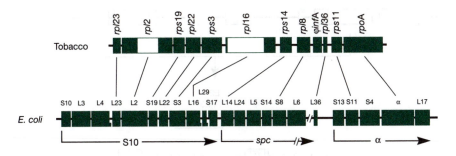
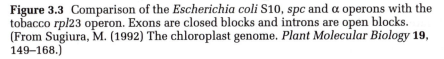

Figure 3.3 Comparison of the *Escherichia coli* S10, *spc* and α operons with the tobacco *rpl*23 operon. Exons are closed blocks and introns are open blocks. (From Sugiura, M. (1992) The chloroplast genome. *Plant Molecular Biology* **19**, 149–168.)

Table 3.2 Higher plant chloroplast genes containing introns

Gene	No. of introns	Function	Remarks
*trn*L-UAA	1	tRNA	
*trn*I-GAU	1	tRNA	
*trn*A-UGC	1	tRNA	
*trn*V-UAC	1	tRNA	
*trn*G-UCC	1	tRNA	
*trn*K-UUU	1	tRNA	
*rps*12	3 exons	Ribosomal proteins	*Trans*-splicing
*rps*16	1	Ribosomal proteins	
*rpl*2	1	Ribosomal proteins	No intron in spinach
*rpl*16	1	Ribosomal proteins	
*rpo*C1	1	Ribosomal proteins	No intron in rice and maize
*clp*P	2	ATP-dependent protease catalytic subunit	No intron in rice and wheat
*pet*B	1	Cytochrome *b/f* complex cytochrome b_6	
*pet*D	1	Cytochrome *b/f* complex subunit IV	
*atp*F	1	Subunit of ATP synthase	
*ndh*A	1	NADH dehydrogenase subunit	
*ndh*B	1	NADH dehydrogenase subunit	
23Sr DNA	1	23S rRNA	

The protein-encoding genes in Table 3.2 and the genes *trn*V, *trn*G and *trn*K have introns with conserved boundary sequences (GTGYGRY at the 5' ends and RYCNAYY(Y)YNAY at the 3' ends, where N is any base, R is a purine and Y is a pyrimidine) and these are sometimes called Group III introns.

The phenomenon of *trans*-splicing is found in chloroplast genomes (see also section 4.2 for *trans*-splicing in mitochondria). Figure 3.4 shows the maturation of the mRNA for four tobacco chloroplast genes *rps*12, *rps*7, *clp*P and

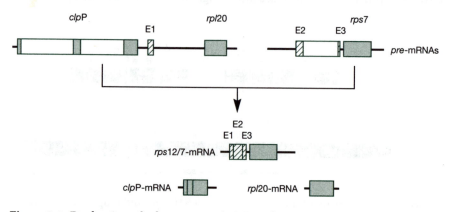

Figure 3.4 Production of tobacco *rps*12/7 mRNA by *trans*-splicing from separate pre-mRNAs. E1, exon 1 of *rps*12; E2 and E3, exon 2 and 3 of *rps*12; open blocks, introns; closed blocks, exons.

*rpl*20 from two primary transcripts. Transcript A contains the gene *clp*P (with two introns), the first exon of *rps*12 and the gene *rpl*20. Transcript B contains exon 2 and 3 of *rps*12 separated by an intron together with the gene *rps*7. *Cis*-splicing removes the two introns from *clp*P (transcript A) and the intron between exon 2 and 3 of *rps*12 (transcript B). However a *trans*-splicing event removes the first exon of *rps*12 from transcript A and splices this on to the other two exons of this gene found on transcript B. This process also produces two monocistronic messages for *clp*P and *rpl*20 and a dicistronic message for *rps*12 and *rps*7.

3.3 Inheritance of chloroplast genes

Chloroplasts are maternally inherited, that is there is essentially no transmission of chloroplasts through the male pollen gamete. This uniparental inheritance of chloroplast genes was first demonstrated in 1909 by Correns in the 4 o'clock plant, *Mirabilis jalapa*. A variegated cultivar of *M. jalapa* called Albomaculata produces leaves with yellowish white patches. The Albomaculata cultivar contains two types of chloroplast: normal chloroplasts that contain chlorophyll and are green, and mutant, uncoloured, chloroplasts (leucoplasts). The latter contain a mutation in a chloroplast gene, which in turn leads to a failure in chlorophyll biosynthesis and therefore no green colour. Many leaf primordial cells will contain a mixture of chloroplasts and this leads to a variegated leaf as, by chance, some leaf cells are produced that contain only leucoplasts. Occasionally shoots are formed from meristematic cells that contain only leucoplasts, forming an all yellow–white shoot, or from meristematic cells that contain only normal chloroplasts, giving an all green shoot.

Flowers produced on these single-colour shoots can be used in controlled crosses to investigate the inheritance of leaf colour and hence chloroplast type. Table 3.3 shows the results of crosses made by Correns. Only one type of progeny was produced in each cross and in each case the progeny resemble the shoot bearing the flowers that act as the female parent, demonstrating uniparental maternal inheritance of chloroplasts. The phenotype (colour) of the shoots bearing flowers used as the source of pollen (the male parent) has no influence on the phenotype of the progeny. This chloroplast mutation is lethal and the yellow–white seedlings produced in these crosses germinate but do not develop fully.

3.4 Control of chloroplast gene expression

Control of transcription

The promoter regions of some chloroplast genes have been studied experimentally using deleted or mutated genes and an *in vitro* transcription system. The promoters of six genes are shown in Figure 3.5; these reveal '–35' and '–10'

Table 3.3 Results of crosses of variegated *Mirabilis jalapa* cv. albomaculata

Phenotype of branch bearing the female parent	Phenotype of branch bearing the male parent	Phenotype of progeny
Yellow–white	Yellow–white	Yellow–white
Yellow–white	Green	Yellow–white
Green	Yellow–white	Green
Green	Green	Green

sequences that are similar to *E. coli* promoter motifs. In addition a sequence motif similar to the nuclear TATA box lies between these two prokaryote-like promoter motifs.

Most chloroplast genes are transcribed polycistronically so that although over 120 genes are present these can be grouped into about 50 transcription units. Multiple transcripts are observed for most chloroplast gene clusters and these are mainly the result of different RNA processing events.

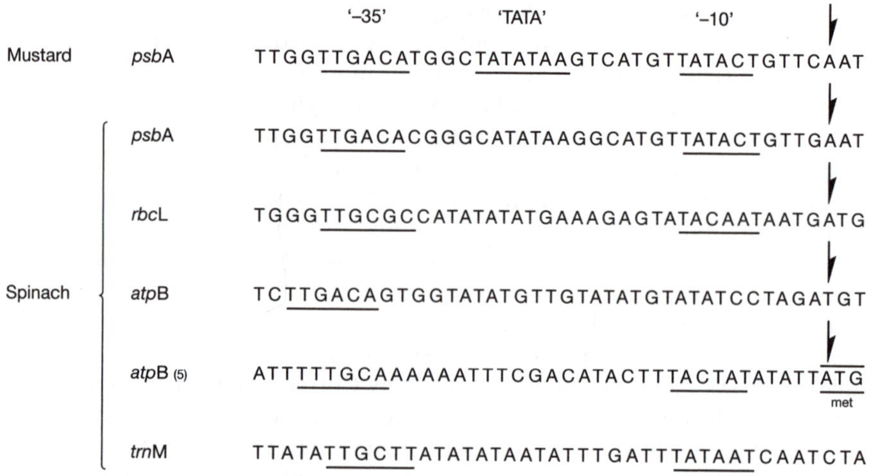

Figure 3.5 Chloroplast promoter regions, identified by using deleted genes and *in vitro* transcription systems. (From Sugiura, M. (1992) The chloroplast genome. *Plant Molecular Biology*, **19**, 149–168.)

Box 3.2 DNA copy number and level of transcription

General transcription activity in chloroplasts is not mediated by fluctuations in DNA copy number. Transcription activity per unit chloroplast DNA and relative chloroplast DNA levels both change considerably during plant development. For example, during spinach leaf maturation (following leaf expansion) the levels of chloroplast DNA double relative to the nuclear DNA content but the transcription rate per unit chloroplast DNA decreases approximately five-fold.

Post-transcriptional control

Some chloroplast genes are known to be constitutively transcribed even though transcript levels change. This suggests that post-transcriptional processing of primary transcripts is an important step in the control of chloroplast genes.

The accumulation of gene transcripts to different steady-state levels during light-induced chloroplast development has been investigated in a number of higher plant species. Figure 3.6 shows the accumulation of mRNA for three chloroplast-encoded genes during the formation of photosynthetically active chloroplasts in spinach cotyledons and leaves. Chloroplast development in the seedling cotyledons is stimulated by 24 hours' light treatment of dark-grown (etiolated) seedlings; in leaves chloroplast development progresses during leaf expansion. The three genes shown in Figure 3.6 are *psb*A, which encodes the 32×10^3 M_r Q_B-binding protein of photosystem II, the ribulose 1,5-bisphosphate carboxylase large subunit gene (*rbc*L) and *atp*B/E encoding the β/ε subunits of ATP synthase.

Steady-state mRNA levels for the three genes were measured by northern blot analysis (Figure 3.6, transcript accumulation). This shows that there are increases in the levels of all three mRNAs during chloroplast development with very dramatic increases in *psb*A and *atp*B/E during cotyledon greening. In contrast, the relative transcription rates of these three genes, as measured by plastid run-on transcription assay, does not change significantly during chloroplast

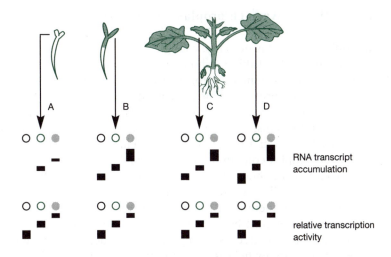

Figure 3.6 Transcriptional and post-transcriptional control of chloroplast gene expression during spinach seedling development. A, 3-day-old seedling with yellow cotyledons; B, 4-day-old seedling with green cotyledons following 24 hours' light; C,D, young and mature leaves from a young plant. The RNA transcript (mRNA) accumulation of three chloroplast genes, *psb*A (O), *rbc*L (O) and *atp*B/E, (●), is measured by northern blotting and the relative transcription activity (transcription rate) is measured by plastid run-on transcription assays. (Adapted from Gruissen, W., Barkan, A., Deng, X-W. and Stern, D. (1988) Transcriptional and post-transcriptional control of plastid mRNA levels in higher plants. *Trends in Genetics*, **4**, 258–263.)

development (Figure 3.6) and the small changes in transcriptional activity observed are clearly not enough to account for the change in steady-state mRNA levels. This discrepancy between the accumulation of transcripts (mRNA) and the rate of transcription (mRNA synthesis) is most clearly seen in *psb*A and *atp*B/E in Figure 3.6. This phenomenon has also been observed in barley for the 16S rRNA and *psa*A–*psa*B (components of photosystem I). These studies suggest that adjustments of plastid RNA stabilities occur and that these play a major regulatory role in chloroplast development and function.

Box 3.3 Prokaryote features of the chloroplast genome

1. Circular double-stranded DNA molecule.
2. DNA not associated with histones.
3. G/C content similar to that of bacteria (36–40%).
4. No methylcytosine.
5. Prokaryote motifs in promoters.
6. Polycistronic mRNA.
7. 70S ribosomes.
8. Protein synthesis inhibited by the antibiotic chloramphenicol.
9. Protein synthesis starts with N-formylmethionine.

3.5 Interaction between chloroplast and nucleus

The chloroplast genome is not large enough to encode all of the proteins present inside the organelle. Table 3.4 shows the individual protein components of two chloroplast protein complexes, together with the location of the genes that encode the individual proteins. This table illustrates that all the major thylakoid membrane protein complexes involved in photosynthesis contain nuclear genome-encoded proteins. Thus, the ATP synthase complex, photosystem I reaction centre (shown in Table 3.4), photosystem II reaction centre and the cytochrome *b*/*f* complex contain both nuclear- and chloroplast-encoded components. Given that the nuclear genes are present only as small (one to five genes) multigene families, whereas the chloroplast genes may be present as 10 000 copies per cell, there is clearly a vast difference in gene copy number. This implies that there must be strict control of nuclear and chloroplast gene function for the coordinated assembly of these structures. The control of nuclear gene expression by light is an important component in this interaction and will be dealt with in Chapter 10.

Another feature of the nuclear genome-encoded chloroplast proteins is also illustrated in Table 3.4. All of the nuclear-encoded proteins are synthesized in the cytoplasm and must be transported into the chloroplast. Table 3.4 shows that all of these proteins are synthesized as a precursor polypeptide that is larger than the mature protein. Thus ferredoxin-binding protein, PS1-2, of the photosystem I reaction centre is synthesized as a 212 amino acid polypeptide but the mature protein, which has a peripheral thylakoid membrane

Table 3.4 Location of genes encoding the proteins of the photosystem I reaction centre and the ATP synthase complex (number of genes in brackets)

Protein	Gene	Location	Amino acid residues in:	
			Precursor polypeptide	Mature protein
Photosystem I				
P700 ch1-*a* apoprotein 1	*psa*A	Chloroplast		?
P700 ch1-*a* apoprotein 2	*psa*B	Chloroplast		?
PS1-2 ferridoxin-binding	*psa*D	Nuclear (1-2)	212	162
PS1-3 plastocyanin-binding	*psa*F	Nuclear (2-3)	231	154
PS1-4	*psa*E	Nuclear (2)	125	91
PS1-5	*psa*G	Nuclear (3-5)	167	98
PS1-6	*psa*H	Nuclear (2-3)	143	95
PS1-7	*psa*C	Chloroplast		81
PS1-8	*psa*I	Chloroplast		36
PS1-9	*psa*J	Chloroplast		45
PS1-10	*psa*K	Nuclear (?)	126	95
PS1-11	*psa*L	Nuclear (?)	209	169
PS1-12	*psa*M	?		
ATP synthase				
CF1-α	*atp*A	Chloroplast		?
CF1-β	*atp*B	Chloroplast		498
CF1-γ	*atp*C	Nuclear (2)	365	323
CF1-δ	*atp*D	Nuclear (1)	257	187
CF1-ε	*atp*E	Chloroplast		153
CFo-I	atpF	Chloroplast		201
CFo-II	*atp*G	Nuclear (1)	222	147
CFo-III	*atp*H	Chloroplast		81
CFo-IV	*atp*I	Chloroplast		247

location, is only 162 amino acids long. The explanation for this difference is that the first 50 N-terminal amino acids of the nascent PS1-2 polypeptide constitute a signal responsible for the correct localization of this protein within the cell and these amino acids are cleaved from the protein during transport across the chloroplast membranes. The signal peptide is responsible both for sorting proteins to the chloroplasts and for transport across the chloroplast membrane.

There are two internal compartments within the chloroplast: the stroma and the lumen of the thylakoid membrane system. Proteins encoded by nuclear genes and synthesized in the cytoplasm but located within the lumen of the thylakoids must be transported across two membranes. These proteins have two signal peptides; the first is removed during transport across the chloroplast envelope membranes by a stromal processing peptidase. A second signal peptide is now at the N-terminus of the polypeptide. This serves to locate the protein to the thylakoid membrane and is removed during transport into the thylakoid membrane lumen by a thylakoid processing peptidase (Figure 3.7).

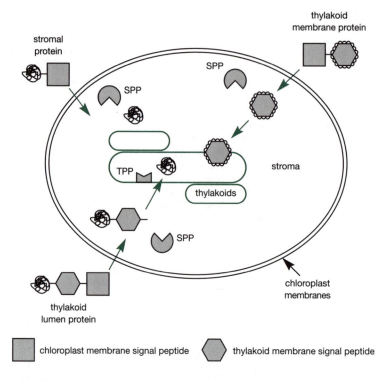

Figure 3.7 Transport of proteins synthesized in the cytoplasm into various compartments within the chloroplast. SPP, stromal processing peptidase; TPP, thylakoid processing peptidase.

Ribulose 1,5-bisphosphate carboxylase/oxygenase

The complex interaction between the chloroplast and the nuclear genome will be illustrated by following the production of the active form of ribulose 1,5-bisphosphate carboxylase/ oxygenase (Rubisco).

Photosynthesis can be simplified into two reactions (Figure 3.8a). The light reaction, which occurs in thylakoid membranes, harvests energy from sunlight to produce ATP and NADPH. These are used in the dark reaction to fix carbon dioxide (CO_2) leading to the production of sugars, amino acids and fatty acids. The dark reaction occurs in the chloroplast stroma. The enzyme, Rubisco, is located in the stroma and is responsible for CO_2 fixation. The reactions catalysed by Rubisco are shown in Figure 3.8b. In relatively high CO_2 concentrations Rubisco will catalyse the formation of two molecules of phosphoglyceric acid from one molecule of ribulose bisphosphate and CO_2. Thus two 3-carbon compounds are formed from one 5-carbon compound and CO_2; that is, net CO_2 fixation. However, if CO_2 levels fall and oxygen (O_2) levels are high, the enzyme has oxygenase activity that can produce one molecule of phosphoglyceric acid (a 3-carbon compound) and one molecule of phosphoglycolic acid (a 2-carbon compound)

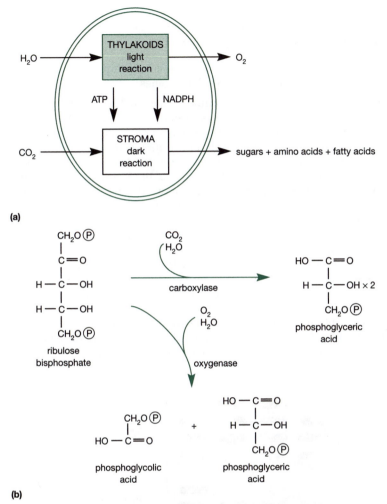

(a)

(b)

Figure 3.8 Biochemical activity of ribulose 1,5-bisphosphate carboxylase/oxyge-
nase (Rubisco). (a) Diagrammatic summary of the light and dark reactions of
photosynthesis. (b) Carboxylase and oxygenase activity of Rubisco.

from one molecule of ribulose bisphosphate with no net CO_2 fixation. This
reaction is the basis of photorespiration.

Rubisco is present at very high levels in the chloroplasts of green leaves
(4–10 mmol l^{-1}) and it has been estimated that it is the most abundant protein
on earth! The active enzyme consists of eight copies of a large subunit (L) of
about 477 amino acids (the exact number depends on the plant species) and
eight copies of a small subunit (S) of about 123 amino acids. It is therefore a
hexadecamer with the formula L_8S_8. The large subunit is encoded by a single
chloroplast gene whereas the small subunit is encoded by a small nuclear
multigene family of 6–12 genes per nuclear genome. In general, not all of the S

subunit genes are active and many species have what are termed pseudogenes, which are not expressed. It has also been shown that different S subunit genes are active at different stages of plant development.

Box 3.4 Location of Rubisco L and S subunit genes

Evidence that the two Rubisco subunits are encoded by genes in different genomes comes from studies of the proteins synthesized in the F_1 hybrid plants produced by reciprocal crosses between *Nicotiana tabacum* and *N. glauca*. *N. glauca* produces three isoforms of the L subunit and a single isoform of the S subunit. *N. tabacum* also produces three isoforms of the L subunit but two isoforms of the S subunit. One of the S subunits of *N. tabacum* and one of the L subunits are different in the two species. The isoforms have different charge properties (isoelectric points) and can be separated with isoelectric focusing gel electrophoresis (Figure a).

Figure a shows the Rubisco L and S subunit isoforms of *N. glauca* and *N. tabacum* together with the isozymes found in the reciprocal F_1 hybrids produced by crossing these two species. When *N. glauca* is used as the female parent the L subunits are identical to *N. glauca*; however, when *N. tabacum* is the female parent the L subunits resemble the *N. tabacum* parent. Thus the Rubisco L subunit shows uniparental maternal inheritance, typical of cytoplasmic (chloroplast) genes. However, the Rubisco S subunits found in F_1 hybrids show biparental inheritance, with uniform amounts of the lower isoform but reduced dosage of the second (higher) isoform.

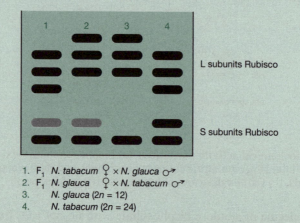

1. F_1 *N. tabacum* ♀ × *N. glauca* ♂
2. F_1 *N. glauca* ♀ × *N. tabacum* ♂
3.   *N. glauca* (2n = 12)
4.   *N. tabacum* (2n = 24)

Figure a Location of Rubisco L and S subunit genes: Coomassie blue-stained isoelectric focusing gel showing the large (L) and small (S) subunit isoforms from *Nicotiana tabacum*, *N.glauca* and the reciprocal F_1 hybrids.

The processes involved in active Rubisco formation are summarized in Figure 3.9. Transcription of the nuclear-encoded S subunit is controlled by light (see Chapter 10) and the resulting mRNA is translated in the cytoplasm on 80S ribosomes to produce a 20×10^3 M_r precursor polypeptide. The N-terminal 46–57 amino acids of the precursor form a transit peptide that is

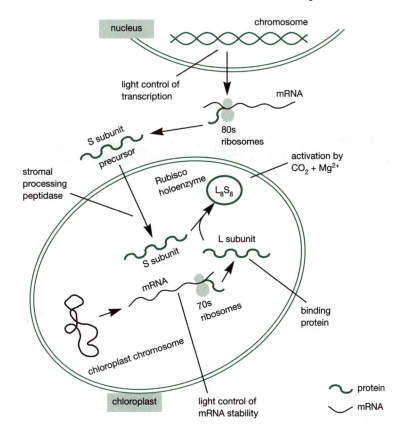

Figure 3.9 Summary of the processes involved in the production of active Rubisco.

responsible for locating the protein in the chloroplasts and for transport of the protein across the chloroplast membranes. The transport process requires ATP and a stromal processing peptidase, which cleaves the transit peptide from the precursor. The stromal processing peptidase itself is known to be nuclear encoded. The L subunit mRNA is transcribed from a chloroplast gene and light controls the stability of this mRNA (see Figure 3.6).

Correct assembly of the L_8S_8 Rubisco molecule requires another protein, the binding protein, which prevents the formation of L subunit aggregates. The binding protein belongs to a class of proteins known as molecular chaperones (chaperonins), which have been shown to be involved in correct protein folding and assembly in a variety of different systems. The binding protein is also nuclear encoded and must therefore be transported into the chloroplast. The correctly assembled Rubisco holoenzyme is only catalytically competent when its active site is complexed with CO_2 and Mg^{2+}. This active ternary complex is formed by a reversible attachment of an activator CO_2 ($^ACO_2$) to the ε amino group of lysine-201 to form a carbamate. This $^ACO_2$ is quite distinct from the substrate CO_2. The slow carbamylation is followed by a rapid binding of Mg^{2+}

to form the active enzyme. In summary, therefore, the production of the active Rubisco L_8S_8 holoenzyme depends upon transcriptional and post-transcriptional controls of gene expression by light, together with transport of proteins (including a chaperonin) into the chloroplast and post-translational activation of the enzyme. However, despite our knowledge of these processes, mechanisms of direct coordination of nuclear and chloroplast gene expression are not known.

Major Learning Objectives for Chapter 3

1. Knowledge of the structure and organization of the chloroplast genome.
2. Understand uniparental inheritance of cytoplasmic genes.
3. Distinguish transcriptional and post-transcriptional controls of chloroplast gene expression.
4. Be aware of the complexity of the interaction between the chloroplast and the nucleus from the example of ribulose 1,5-bisphosphate carboxylase/ oxygenase (Rubisco).

Further reading

BRYCE, J.H. and HILL, S.A. (1993). Energy production in plant cells. In Lea, P.J. and Leegood, R.S. (eds) *Plant biochemistry and molecular biology*, pp. 1–26. John Wiley & Sons, Chichester.
Contains an overview of the biochemistry of photosynthesis.

GRUISSEN, W., BARKAN, A., DENG, X.W. and STERN, D. (1988). Transcriptional and post-transcriptional control of plastid mRNA levels in higher plants. *Trends in Genetics*, **4**, 258–263.
A clear and concise account of the control of chloroplast gene expression.

SUGIURA, M. (1992). The chloroplast genome. *Plant Molecular Biology*, **19**, 149–168.
A fuller and more recent account of the chloroplast chromosome than Umesono and Ozeki (1987); also contains information about chloroplast genes and their expression.

UMESONO, K. & OZEKI, H. (1987). Chloroplast gene organisation in plants. *Trends in Genetics*, **3**, 281–287.
A concise account of the chloroplast chromosome; contains a map of the *Marchantia polymorpha* chloroplast genome.

Chapter 4

Mitochondria

4.1 Mitochondrial genome organization

The respiration of carbon compounds occurs within cell organelles called mitochondria. The biochemistry of respiration will not be covered here but a
reference covering this topic is given at the end of the chapter. Like chloroplasts,
mitochondria are replicative organelles and contain their own chromosome.

Plant mitochondrial genomes are larger than the mitochondrial genomes of
mammals and yeast. However, the size varies greatly between plant species.
Based on restriction enzyme and reannealing analysis *Brassica campestris* has a
mitochondrial genome of 218 kb but muskmelon, *Cucumis melo*, has a mitochondrial genome of 2400 kb.

Box 4.1 Mitochondrial genome size

1. The variation in mitochondrial genome size between species of plant
   cannot be explained by large differences in the amount of repeated DNA
   which, at the most, only represents 10% of the genome.
2. The larger genomes are unlikely to be due to a much larger number of
   genes in some mitochondria, since studies of protein synthesis by isolated
   mitochondria indicate a similar number of mitochondrial polypeptides
   among plant species having great differences in genome size.
3. The size and structure of the mitochondrial genes is also unlikely to
   account for the interspecific mitochondrial DNA (mtDNA) size variation,
   even though mitochondrial genes can contain introns (see section 4.2).
4. Although, in every plant species examined, mtDNA contains integrated
   chloroplast DNA (ctDNA), sometimes called promiscuous DNA, there is
   no evidence that the large mitochondrial genomes contain more ctDNA,
   for example the total amount of ctDNA in the maize mtDNA is 4–5%.

Plant mitochondrial genome structure has proved difficult to analyse partly
because it can undergo rapid changes in structure. The small, circular mitochondrial (mt) DNA molecules of animal cells (~16 kb) can be visualized by
electron microscopy; by extrapolating from these molecules, and the fact that

overlapping restriction fragments of plant mtDNA can be arranged to form circular maps, a circular structure has been postulated for the large plant mitochondrial genomes. However, most plant mtDNA behaves in gel electrophoresis like linear molecules, ranging from 50 to 100 kb, in a heterogeneous population visible as a smear. The linear molecule interpretation could explain the difficulty in identifying large circular molecules by electron microscopy. Subgenomic pieces of the plant mitochondrial genome can be found as circles but which are too small to be the complete genome. A tripartite circular model has been proposed for the *B. campestris* genome (Figure 4.1) but the precise physical form of the 'master' mitochondrial chromosome is not known.

The subgenomic mtDNA circles are proposed to arise from frequent recombination, at sites containing small nucleotide repeats (Table 4.1). The maize 'N' mitochondrial genome is 570 kb (540 kb of unique sequence) with seven large (>700 bp) repeated sequences and many short repeats. It has been suggested that the large repeats may arise as a consequence of recombination between small repeats and the small repeats may be derived from reverse transcription of the processed 'non-functional' parts of mitochondrial transcripts (see section 4.2).

Box 4.2 *Marchantia polymorpha* mitochondria

In 1992 the entire genomic sequence of the mitochondrial DNA of the simple plant *M. polymorpha* (a liverwort) was published and this has greatly advanced our knowledge of the information content of plant mitochondrial genomes.

There are 94 substantial open reading frames in the *M. polymorpha* mtDNA, which is about seven times the number found in animal mtDNA (13 open reading frames). A large proportion of the genome consists of noncoding sequences between the genes. Only the *nad7* locus and several species-specific cytoplasmic male sterility-related reading frames identified in higher plant mtDNA are absent from M. polymorpha mtDNA. Unlike higher plants there are no ctDNA sequences in the liverwort mtDNA, and the lack of integrated ctDNA in M. polymorpha mtDNA may be related to the fact that it seems not to undergo extensive recombination.

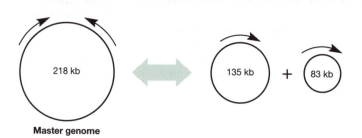

Master genome

Figure 4.1 Tripartite model of *Brassica campestris* mitochondrial chromosome structure: a large 'master' chromosome of 218 kb and two small chromosomes (135 kb and 83 kb), which arise by homologous recombination at the large repeats (black arrows) on the 'master' chromosome.

Table 4.1 Small repeats associated with recombination in plant mtDNA

Plant	Gene/sequence	Repeat size (bp)
Maize	*cms*-T revertants	127
		58
	NCS 3 mutant	12
	NCS 5 mutant	6
	RU → N duplication	181
	(two repeats)	241
Brassica (Ogura radish)	*atp*6 gene	118
		241
	*atp*1 gene	689
	*cox*I sublimon	689
Wheat	tRNApro gene duplication (several repeats)	7–188
Sunflower	*atp*1 gene from cms	262

There can be up to seven-fold variation in the amount of DNA of individual regions of the mitochondrial genome present in mitochondria from a single plant. This implies that there must be differential replication of subgenomic sequences and very few copies of the 'master' genome, which in *B. campestris* is present as three copies per organelle.

Almost all of the genes for subunits of the respiratory chain protein complexes identified in animal mtDNA have been found in plant mtDNA (Table 4.2). Similar to animal and yeast mitochondria these active membrane complexes contain proteins encoded by both mitochondrial genes and nuclear genes.

Movement of genetic information, on an evolutionary scale, from mitochondria to the nucleus has been deduced for several higher plant species. Examples include a mitochondrial intron fragment upstream of a nuclear lectin gene in an African bean; an active *cox*II gene in pea nuclear DNA; and the *rps*12 gene, which is nuclear in *Oenothera berteriana*. Correct nuclear expression of such integrated mitochondrial genes requires several changes because of the special features of transcript processing in mitochondria (see section 4.2). These can be accomplished most easily by reverse transcription of the mitochondrial mRNA followed by the addition of the correct signals for nuclear transcription and cytoplasmic translation.

4.2 Mitochondrial gene expression

Transcript processing

The protein products of plant mitochondrial genes cannot be predicted accurately from genomic sequences since RNA editing modifies virtually all mRNA sequences

Table 4.2 Genes identified in plant mitochondrial genomes

Function/product	Genes
Translation apparatus	
Ribosomal RNA	*rrn*5, *rrn*18, *rrn*26
Ribosomal protein[a]	
Small subunit	*rps*1, *rps*3, *rps*7, *rps*12
	*rps*13, *rps*14, *rps*19
Large subunit	*rp*12, *rp*15, *rp*116
Transfer RNA	At least 16
Subunits of respiratory chain complexes	
NADH dehydrogenase	*nad*1, *nad*2, *nad*3, *nad*4, *nad*41, *nad*5, *nad*6, *nad*7, *nad*9
Cytochrome *b*	*cob*
Cytochrome *c* oxidase	*cox*I, *cox*II, *cox*III
ATP synthase	*atp*1, *atp*6, *atp*9
Cytochrome *c* biogenesis	At least 4 genes
Conserved open reading frames[b]	At least 10 known

[a] In *Marchantia polymorpha* there are 16 genes for ribosomal proteins and a similar number may exist in higher plants.
[b] The conservation of these open reading frames suggests an important function, which is not known.

post-transcriptionally. Typically editing involves changes from C bases in the genome to U in the mRNA. RNA editing alters 5' untranslated regions (UTRs), 3' UTRs as well as introns and exons and it may be required for RNA processing.

Several mitochondrial genes contain introns. The mitochondrial genome of the liverwort, *M. polymorpha*, contains 32 self-splicing introns: 25 are members of Group II introns and seven are classified as Group I introns (see Table 4.3 for types of intron). Six of the Group I introns are in the *cox*I gene but the *cox*I genes (encoding subunit 1 of cytochrome oxidase) of higher plants lack introns and no Group I introns have been found in higher plant mtDNA.

> **Box 4.3 *Marchantia polymorpha cob* genes**
>
> Duplicate *cob* genes (cytochrome *b*) exist in *M. polymorpha* mtDNA: one copy of the gene contains introns whereas the other has an uninterrupted coding sequence.

Sequences with homology to viral reverse transcriptase have been identified in a number of mitochondrial introns. The functional significance of this is not

Table 4.3 Intron splicing

Type	*Cis*-splicing mechanism	*Trans*-splicing counterpart
tRNA	Simple, enzymatic 3' cyclic phosphate intermediate	None known
Group I self-splicing	Simple, autocatalytic GTP-dependent	None known
Group II self-splicing	Simple, autocatalytic lariat intermediate	Chloroplast *rps*12 gene, mitochondria genes *nad*1, 2, 4, 5
Nuclear pre-mRNA splicing	Complex, enzymatic lariat intermediate	

understood, although reverse transcription has been implicated in a number of mtDNA phenomena.

In higher plant mitochondria, transcripts of the NADH dehydrogenase complex (*nad1,2,5*) are processed by *trans*-splicing that connects exons scattered throughout the genome (Figure 4.2). The mature transcripts are assembled via split Group II introns (Figure 4.3) so that the sequences can be aligned by base pairing to form the typical Group II intron structure. This *trans*-splicing mechanism appears to predate the divergence of monocots and dicots. *Cis*-spliced introns also exist in the NADH dehydrogenase genes (Table 4.4).

Translation

The mitochondrial genome encodes part of the mitochondrial protein translation machinery (Table 4.2). However, not all the required tRNA genes are present in higher plant mtDNA and additional tRNAs have to be imported into mitochondria from the cytoplasm.

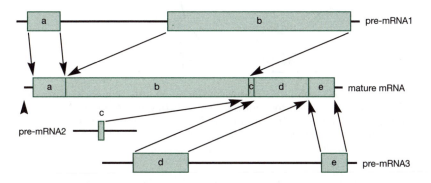

Figure 4.2 *Trans*-splicing: assembly of the mature mRNA for the *nad*5 mitochondrial gene from three different pre-mRNA molecules involves both *trans* and *cis* splicing (this organization has been found in *Arabidopsis*, wheat and maize). (Adapted from Wissinger, B., Brennicke, A. and Schuster, W. (1992) Regenerating good sense. *Trends in Genetics*, **8**, 322.)

Table 4.4 Examples of *cis-* and *trans*-splicing introns in higher plant mitochondria

Gene	Cis introns	Trans introns	Intron editing
*nad*1			
(wheat/*Petunia*)	1	3	?
(*Oenothera/Arabidopsis*)	2	2	+/-
nad2 (*Oenothera*)	3	1	+
nad4 (wheat/*Arabidopsis*)	3	—	?
nad5 (*Oenothera/Arabidopsis*)	2	2	+

—, No intron of the respective type present;
+, intron editing identified; ?, intron sequences not investigated.

The ribosomes of mitochondria are structurally more like prokaryote ribosomes. They have a sedimentation coefficient of 77–78S and unlike cytoplasmic ribosomes (80S) are inhibited by the antibiotics chloramphenicol and tetracycline. Plant mitochondrial ribosomes contain a distinctive 5S rRNA not found in other eukaryote mitochondrial ribosomes. Both rRNAs and tRNAs of plant mitochondria are more like the rRNAs and tRNAs from chloroplasts and eubacteria than the same molecule in animal and yeast mitochondria.

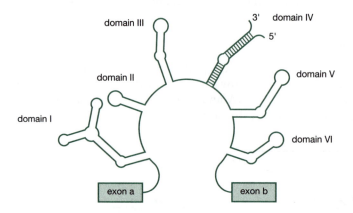

Figure 4.3 Model of secondary structure of the nucleotide sequences flanking the *trans*-spliced exons in mitochondria. The ends of the two pre-mRNA molecules are shown base paired in domain IV. This structure is similar to that of Group II introns. (Adapted from Wissinger, B., Brennicke, A. and Schuster, W. (1992) Regenerating good sense. *Trends in Genetics*, **8**, 322.)

Like chloroplast mRNAs, those from mitochondria lack the 5' cap and usually lack a 3' poly(A) tail. Mitochondria also use a slightly different genetic code, e.g. the codon CUA codes for leucine in nuclear genes but codes for threonine in mitochondria (yeast) and the nuclear 'stop' codon UGA codes for tryptophan in mitochondria.

4.3 Cytoplasmic male sterility (CMS)

A maternally inherited male sterile phenotype is known in many plant species, including maize and *Petunia hybrida*. The male sterile phenotype arises from a failure in pollen development in the anthers. Since all seed produced on the male sterile parent must be hybrid in origin, this character has been the subject of considerable research because of the commercial interest in the production of F_1 hybrid seed. Both chloroplast and mitochondrial DNAs are maternally inherited in higher plants and the initial search for the location of the CMS genetic determinant was therefore in organelle DNA. In addition to mtDNA, CMS has been found to depend on nuclear genes, known as fertility restorers (*Rf*). As the term implies, these nuclear genes modify the effect of the mtDNA CMS factor to allow the production of normal pollen.

Chimeric structure of Petunia CMS-associated DNA

An mtDNA region, which includes an RFLP marker, was found to be associated with CMS in *Petunia*. This CMS-associated DNA (*S-pcf*) has been sequenced and found to be a mosaic or chimeric structure containing sequences from five different genes (Figure 4.4). Transcripts and a protein product of the *S-pcf* gene have been identified in male sterile plants. There is evidence to suggest that the nascent *S-pcf* encoded polypeptide is post-translationally processed but a function for the protein has not been determined.

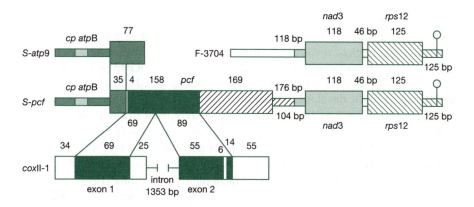

Figure 4.4 Mitochondrial DNA associated with cytoplasmic male sterility (CMS) in *Petunia* (the *S-pcf* locus) and homologous mitochondrial genes, showing the chimeric nature of this locus. Large boxes indicate exons, narrow boxes indicate flanking regions. The number of codons is indicated above each exon; the size of flanking regions is indicated in base pairs (bp). (From Hanson, M.R., Connett, M.B., Folkerts, O., Izhar, S., McEvoy, S.M., Niivison, H.T. and Pruitt, K.D. (1991) Cytoplasmic male sterility in *Petunia*. In Herrmann, R.G. and Larkins, B.A. (eds) *Plant molecular biology*, vol. 2, pp. 383–399. Plenum Press, New York.)

Maize CMS

In maize, there are three types of CMS (*cms*-T, *cms*-C and *cms*-S); these are characterized by the unique nuclear restorer genes (*Rf*), which suppress each type of CMS. Thus genes *Rf*1 and *Rf*2 restore fertility to *cms*-T but not *cms*-C or *cms*-S. They are also distinguished by characteristic differences in mtDNA restriction profiles and by a characteristic set of polypeptides synthesized by isolated mitochondria. In *cms*-T mtDNA an open reading frame encoding a 13×10^3 M_r protein (URF13) produces a transcript only found in male sterile plants. URF13 is also produced by a remarkable chimeric gene (T-*urf*13) apparently originating from multiple rearrangements that have assembled the flanking and/or coding regions of the mitochondrial 26S rRNA gene (*rrn*26), the *atp*6 gene and a promiscuous chloroplast tRNA*arg* gene. Like *S-pcf* in *Petunia*, T-*urf*13 is a protein-encoding gene and therefore it is surprising that its coding region consists mostly of sequences from the coding and flanking regions of *rrn*26, a structural RNA gene. The 13×10^3 M_r URF13 polypeptide is a component of the inner mitochondrial membrane (Figure 4.5).

Box 4.4 Membrane location of URF13

DCCD (dicyclohexylcarbodiimide) is a protein-modifying reagent that forms a stable covalent adduct with acidic amino acid side-chains located in hydrophobic domains. The Asp-39 of URF13 forms such a stable compound with DCCD, which locates this residue within the membrane. Figure 4.5 shows a model for the membrane-located structure of URF13.

Evidence linking the T-*urf*13 gene product with CMS comes from an alteration in the T-*urf*13 transcript in reverse mutations and in plants containing the functional nuclear restorer genes. Although it is not known how *Rf*1 operates, it alters the transcription of T-*urf*13 and reduces URF13 protein level by about 80%. The T-*urf*13 transcript appears to be expressed constitutively and is not restricted to the developing anthers.

During the 1950s and 1960s, *cms*-T was widely used in the production of hybrid maize. In 1969 and 1970 *cms*-T cytoplasm was present in 85% of maize acreage in the USA – in the southern states and the 'corn belt' the crop was severely affected by corn blight caused by the fungus *Bipolaris maydis* race T (formerly *Helminthosporium maydis*). Maize plants containing *cms*-C or *cms*-S cytoplasm are only mildly affected by this pathogen. Another fungal pathogen, *Phyllostica maydis*, is also uniquely virulent on *cms*-T cytoplasm. The susceptibility of *cms*-T to *B. maydis* race T is caused by mitochondrial sensitivity to a fungal toxin (BmT toxin) that has been shown to cause rapid permeabilization of the inner mitochondrial membrane.

The simplest model to explain how URF13 renders the membrane sensitive to BmT toxin involves a direct interaction between URF13 and the toxin. The current model for protein structures that form hydrophilic pores through membranes involves the association of membrane-spanning amphipathic α-helices in a cylindrical structure such that the polar surfaces of the helices face inwards. Because NAD⁺ can pass through pores generated by BmT toxin interaction with URF13, the channel must be 1.5 nm diameter and this requires an oligomeric URF13 structure (Figure 4.6).

CMS and BmT toxin sensitivity appear to be inseparable and T-*urf*13 is believed to be responsible for both traits. It is suggested that an anther-specific substance is produced that affects mitochondria in a manner similar to BmT toxin causing permeabilization of the inner membrane and loss of mitochondrial activity. This in turn would lead to pollen abortion. An anther-specific substance of this nature has not yet been identified.

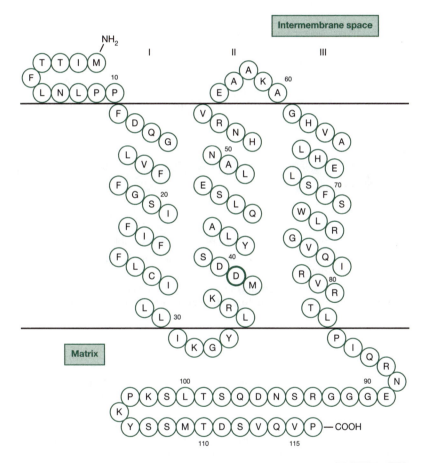

Figure 4.5 Model of the membrane location of the $13 \times 10^3 M_r$ polypeptide (URF13) associated with cytoplasmic male sterility (*cms*-T) in maize, showing the putative membrane-spanning helical domains. Amino acids shown in the single letter code. (From Levings, C.S. and Siedow, J.N. (1992) Molecular basis of disease susceptibility in the Texas cytoplasm of maize. *Plant Molecular Biology*, **19**, 135–147.)

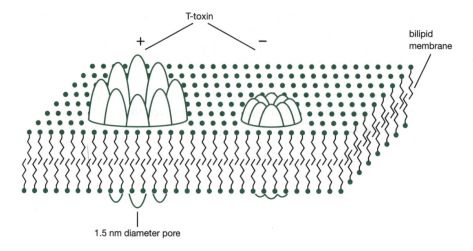

Figure 4.6 Structure of the proposed oligomeric URF13 transmembrane complex, showing the change in conformation in the presence of fungal T-toxin, producing 1.5-nm pores. (From Levings, C.S. and Siedow, J.N. (1992) Molecular basis of disease susceptibility in the Texas cytoplasm of maize. *Plant Molecular Biology*, **19**, 135–147.)

Major Learning Objectives for Chapter 4

1. Knowledge of the organization and information content of plant mitochondrial genomes.
2. Be acquainted with the instability of plant mitochondrial genomes.
3. Understand mitochondrial transcript processing and the unique features of the mitochondrial genetic code.
4. Understand the molecular basis of cytoplasmic male sterility and the susceptibility of *cms*-T maize to the fungal pathogen, *B. maydis* race T.

Further reading

BRYCE, J.H. and HILL, S.A. (1993). Energy production in plant cells. In Lea, P.J. and Leegood, R.S. (eds) *Plant biochemistry and molecular biology*, John Wiley & Sons, Chichester. pp.1–26.
Contains an overview of the biochemistry of the respiration of carbon compounds.

FAURON, C., CASPER, M., GAO, Y. and MOORE, B. (1995). The maize mitochondrial genome: dynamic, yet functional. *Trends in Genetics*, **11**, 228–235.
An account of the instability of the maize mitochondrial chromosome; contains details of the DNA rearrangements seen in the mitochondria of different maize cultivars.

LEVINGS, C.S. and SIEDOW, J.N. (1992). Molecular basis of disease susceptibility in the Texas cytoplasm of maize. *Plant Molecular Biology*, **19**; 135–147.
An account of CMS in *cms*-T maize and the relationship between CMS and susceptibility to the maize pathogen *B. maydis* race T.

SCHUSTER, W. and BRENNICKE, A. (1994). The plant mitochondrial genome: physical structure, information content, RNA editing and gene migration to nucleus. *Annual Review of Plant Physiology and Plant Molecular Biology*, **45**, 61–78.
A full review of the structure of the mitochondrial genome and mitochondrial genes plus mitochondrial gene expression.

WISSINGER, B., BRENNICKE, A. and SCHUSTER, W. (1992). Regenerating good sense: RNA editing and *trans* splicing in plant mitochondria. *Trends in Genetics*, **8**, 322–328.
A concise explanation of RNA editing and *trans*-splicing in plant mitochondria.

Chapter 5

Transposable elements

5.1 Barbara McClintock and the discovery of maize transposable elements

Barbara McClintock was a maize cytogeneticist who worked for most of her scientific career at the Cold Spring Harbor Laboratories in the USA. She was born in 1902 and by the 1930s had an international reputation as a cytogeneticist. Following her move to Cold Spring Harbor, she started working on a number of maize crosses that showed a range of genetic instabilities. These were identified by chromosomal breakage events and by the production of mosaic (variegation) patterns in somatic tissue of the maize grain. In 1951 (as a lecture) and 1953 (as a paper) her work on these instabilities was published. These papers represented a genetic analysis of what are now called transposable elements. In fact the word transposition was first used by Barbara McClintock in 1949. The genetic analysis was, and still is, difficult to follow and the significance of her work was not understood by the scientific community. In fact, she only received two reprint requests for her 1953 published paper.

Barbara McClintock discovered autonomous and non-autonomous transposable elements in maize, one type of which she gave the symbols Ac (for activator) and Ds (for dissociator). She recognized that the elements could move in the nuclear genome, that their movement caused mutations and that the movement of Ds was dependent on the presence of Ac. It was not until the discovery and molecular studies of transposable elements in bacteria in the 1970s that a molecular explanation of maize transposable elements confirmed Barbara McClintock's analysis. In 1983, 30 years after the publication of her studies on maize transposable elements, she received the Nobel Prize for Medicine for her work.

Transposable elements (or transposons) are pieces of DNA that can move (transpose) within the nuclear genome. Transposons are autonomous if they control their own movement; non-autonomous transposons also exist and these require the presence of another transposable element in the genome for transposition. They often cause insertional mutations by integrating into functional genes and destroying their structural integrity. Subsequent movement of the transposable element out of the mutant gene (and thus restoration of gene function) causes somatic cell reverse mutations. These mutations can often be seen as a mosaic pattern or variegation of the colour or texture of the tissue, if they occur during the development of a tissue in which the gene is active. The Mendelian inheritance of variegation patterns is the first criterion by which transposable elements are recognized. This distinguishes them from variegation caused by viral disease, by chimeras or by chloroplast mutations.

Box 5.1 *Antirrhinum* flower pigmentation and Tam transposons

A wide range of mutant plants which have been noticed and propagated by the horticultural industry because they have a patterned or mosaic phenotype, have been shown to arise from transposon insertional mutations. Figure a shows in outline the biosynthetic pathway for anthocyanin pigment synthesis. The arrows represent one or more enzymatic steps and the position of blocks caused by mutations in the *nivea* and *pallida* genes in snapdragon (*Antirrhinum majus*) are shown as bars. Table a gives the transposon position in three alleles of *nivea* and one of *pallida*. Tam1 and Tam3 are autonomous transposable elements that can move in the genome and give rise to variegated (patterned) flowers. Tam2 is a non-autonomous transposable element that cannot move in the genome by itself, and gives rise to a stable mutant phenotype.

Table a Transposable elements in *Antirrhinum majus*

Allele	Flower phenotype	First documented	Element	Element position[a]
*niv*rec-53	Red sites on white background	1936	Tam1	--▽-■------■----■----■-- ↳ 17 bp upstream of TATA box
niv-44	Stable white	1955	Tam2	----■▽------■---- ■-- ↳ First exon/intron boundary
*niv*rec-98	Red sites on pale background	1979	Tam3	--▽-■------■---- ■-- ↳ 29 bp upstream of TATA box
*pal*rec-2	Red sites on ivory background	1868	Tam3	--▽-■--......... ↳ 41 bp upstream of TATA box

[a] The triangle indicates element position, the arrow shows the site and direction of transcription initiation and the thick bars indicate exons. Only the first exon of the *pal* locus has been sequenced. The overall length shown of the *niv* locus is 2.5 kb.

▶

(Box 5.1 continued)

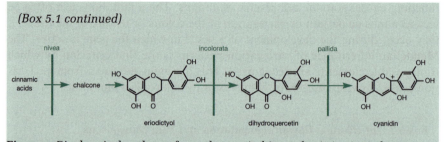

Figure a Biochemical pathway for anthocyanin biosynthesis in *Antirrhinum majus* (snapdragon) showing the positions of blocks caused by the *nivea* and *pallida* mutations. The arrows represent one or more enzyme steps. (From Coen, E.S. and Carpenter, R. (1986) Transposable elements in *Antirrhinum majus*: generators of genetic diversity. *Trends in Genetics*, **2**, 292–296.)

There are two distinct classes of transposable element known and these are distinguished because they transpose via different mechanisms. In retrotransposons the elements move via an RNA intermediate; very few active retrotransposons have been identified in plants. The other class of transposons, of which Ac and Ds (as well as Tam1, 2 and 3) are examples, move via direct excision and integration of DNA sequences. This chapter will concentrate primarily on the analysis and use of the maize Ac and Ds transposable elements discovered by Barbara McClintock, since this system is the most thoroughly studied in plants.

5.2 Ac and Ds transposable elements in maize

Structure

The Ac transposable element is 4563 bp long and has an 11-bp IR at the ends. The transposon encodes a single 3.5-kb transcript that includes a 650-bp 5' UTR and an open reading frame (ORF) encoding an 807 amino acid protein with transposase activity. This transposon gene has four introns dividing the sequence into five exons. Flanking the transposase gene and within the IRs are two regions containing the *cis* determinants for excision. The structure of the Ac transposable element is shown in Figure 5.1.

All of the Ds transposable elements that have been sequenced have the Ac 11-bp IR at the ends. Most of the Ds elements have extensive homology to the Ac element but are shorter, having apparently arisen by deletion of part of the Ac internal ORF. Figure 5.2 shows a comparison of two Ds elements and the full-length Ac element. Since Ds elements do not have the complete ORF, they produce no active transposase and cannot therefore move by themselves. However, these non-autonomous Ds elements can move if Ac is also present in the genome since Ac produces the transposase and this will mobilize the Ds elements. Since the Ac and Ds elements do not have to be linked for this mobilization, Ac is said to work in *trans*. Not all Ds elements have this deletion

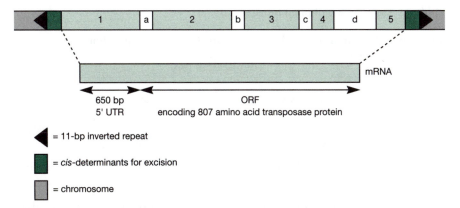

Figure 5.1 Structure of autonomous transposable element, Ac, from maize. ORF, open reading frame; 1–5, exons; a–d, introns.

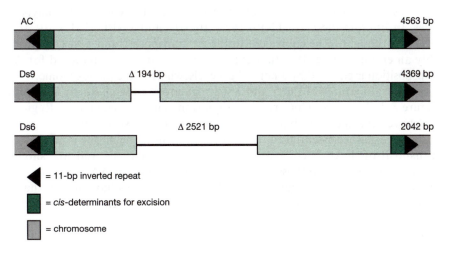

Figure 5.2 Comparison of the structure of the autonomous transposable element Ac with the non-autonomous transposable elements Ds9 and Ds6. Δ, deletion.

structure: Ds2 has the Ac 11-bp repeat but very little homology to Ac; Ds5933, which causes chromosome breakage, has a double structure.

Table 5.1 shows a comparison of the maize (*Zea mays*) Ac transposable element with two types of transposable element from snapdragon (*Antirrhinum majus*) and another, different, maize transposable element system (En/Spm – enhancer/suppressor) isolated by Barbara McClintock. Tam3 from *A. majus* is about the same size as Ac, has a 12-bp IR at the ends and contains a single ORF (with no introns). A region of 520 amino acids of the putative protein encoded by this ORF has 30% identity to the Ac transposase. A further similarity between Ac and Tam3 is that both generate an 8-bp target site duplication when they insert into DNA.

Table 5.1 Comparison of plant transposons

Element	Terminal inverted repeat	Target site duplication (bp)	Size (bp)	Species
Ac	^C_TAGGGATGAAA	8	4563	*Zea mays*
Tam3	TAAAGATGTGAA	8	4869	*Antirrhinum majus*
En/Spm	CACTACAAGAAAA	3	8287	*Zea mays*
Tam1	CACTACAACAAAA	3	15164	*Antirrhinum majus*

The other two transposons shown in Table 5.1 also share structural similarities with each other. En/Spm from maize and Tam1 from *A. majus* both produce a 3-bp target site duplication upon insertion. Although En/Spm and Tam1 differ in size they are both much larger than Ac and Tam3 and in both transposons a single precursor transcript is predicted to encode two polypeptides. In En/Spm these are TNPA (67×10^3 M_r) and TNPD (131×10^3 M_r). TNPA is thought to be equivalent to the Ac transposase and TNPD is probably an endonuclease. It is thought that the endonuclease required for Ac transposition is encoded by a normal gene, elsewhere in the plant genome.

Structural features of the 5' and 3' *cis*-regulatory ends of Ac are shown in Figure 5.3. These are CpG-rich, methylation-sensitive sequences. Regions that have been shown to bind to purified Ac transposase are shown in green and regions either essential for transposition or influencing the efficiency of transposition have been mapped by mutation analysis and are shown hatched. In addition to these features, the sequence motif AAACGG occurs in clusters in these two regulatory ends. Similar studies have shown that a 12-bp motif is also present six times in the 5' and eight times in the 3'

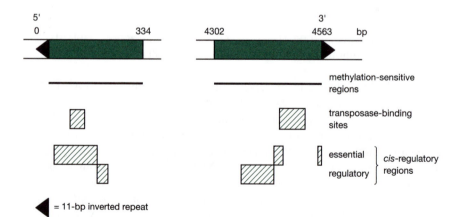

Figure 5.3 Structure and function of the 5' and 3' *cis*-determinant regions for excision of the maize transposable element Ac.

cis-regulatory ends of the En transposon and these motifs are recognized by the TNPA (transposase) protein.

Mechanism of transposition

The mechanism of transposition is not well understood but models for the mechanism have to include:

1. association of two ends of transposable element;
2. endonuclease activity to cut close to the end of the integrated element or at target site;
3. ligation of chromosome ends.

There are a number of features common to the Ac and Spm elements that provide some experimental evidence for these processes:

1. an element encoded protein (transposase) that binds to *cis*-regulatory regions;
2. asymmetry of *cis*-regulatory regions;
3. terminal IRs, as possible recognition sites for an endonuclease;
4. duplication of bases at the target site.

The current model for excision of the transposable element involves binding of transposase to the two *cis*-regulatory regions in order to produce a complex of the two ends and recognition of the terminal IR by the transposon endonuclease in Spm and a 'plant' endonuclease in Ac. During insertion a staggered cut is made, which causes the target site duplication.

5.3 Transposon tagging

A number of transposable elements have been cloned and these have been used to isolate mutant genes from genomic libraries produced from plants containing a homologous transposon insertional mutation, that is containing the same transposon. This strategy used the transposon DNA as a probe for the mutant allele, followed by identification of flanking sequences, which represent the gene, and subsequent use of this DNA to isolate the gene from a wild-type plant genomic library. A list of genes isolated in this way is shown in Table 5.2.

Transposition in a heterologous species

Cloned transposable elements have also been introduced into a number of different (heterologous) plant species (see Chapters 6 and 7 for the method of introducing cloned genes into plant nuclear genomes). Table 5.3 shows some examples of transgenic plants containing Ac, where both autonomous and non-autonomous transposition of Ds (or experimentally produced deletions of Ac) have been observed.

Table 5.2 Genes isolated using transposon tagging in plants

Gene	Gene function	Species	Transposon
*A*1	Enzyme required for anthocyanin biosynthesis	*Z. mays*	En/Spm
*Bz*1	Enzyme required for anthocyanin biosynthesis	*Z. mays*	Ac
*Bz*2	Enzyme required for anthocyanin biosynthesis	*Z. mays*	Ds
*C*1	Regulatory gene	*Z. mays*	En/Spm
*C*2	Enzyme required for anthocyanin biosynthesis	*Z. mays*	En/Spm
P	Regulatory gene	*Z. mays*	Ac
R	Regulatory gene	*Z. mays*	Ac
Knotted	Leaf development	*Z. mays*	Ds
Opaque-2	Regulatory gene	*Z. mays*	Ac; En/Spm
Pallida	Enzyme required for anthocyanin biosynthesis	*A. majus*	Tam3
Deficiens	Regulatory gene	*A. majus*	Tam7

Box 5.2 Experiment to demonstrate transposition in heterologous species

One method for measuring transposition in a heterologous species is illustrated in Figure a. Part (1) of Figure a shows three constructs that contain both the gene for kanamycin resistance (*Npt*II) and the gene for hygromycin resistance (*Hpt*). Plant cells will not grow in the presence of either of these two antibiotics; however introduction of the construct into the nuclear genome of plants will allow the resulting transformed cells to grow in media containing these antibiotics. Two of the constructs shown in Figure a part (1) have a transposable element in the *Npt*II gene and these insertion mutations prevent these constructs conferring kanamycin resistance to the recipient plant cells.

It is possible to take cells from the leaves of a tobacco plant (*N. tabacum*) and use an enzyme cocktail to remove the cell walls from them; the resulting cells become spherical and are known as protoplasts. The protoplasts can be plated out onto agar plates (Petri dishes) containing a suitable medium, where they reform cell walls and divide to form small colonies of cells. Part (2) of Figure a shows the results of an experiment where the three constructs were introduced separately into three aliquots of a tobacco protoplast suspension. The resulting transformed cells were then plated out onto agar plates containing either kanamycin and hygromycin or hygromycin alone. Figure a Part (2) shows that protoplasts containing construct A can grow in

▶

(Box 5.2 continued)

the presence of both antibiotics. Protoplasts containing construct C can only grow on the plate containing hygromycin alone because the presence of the non-autonomous Ds transposon in *Npt*II means that this gene is not expressed and the cells are not resistant to kanamycin. The result of plating protoplasts containing construct B onto plates containing both antibiotics is different from the other two tests, with a small number of cell colonies being formed. This result demonstrates the ability of the Ac transposable element to excise from the *Npt*II gene, thereby restoring gene activity and kanamycin resistance of the cells and allowing the formation of a small number of colonies.

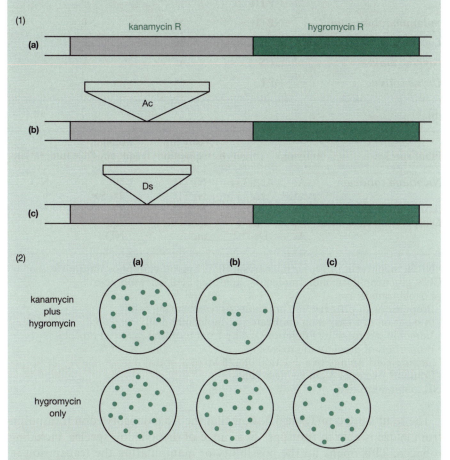

Figure a Experiment to demonstrate the transposition of Ac in a heterologous species. (1) TDNA constructs containing (a) the genes for hygromycin and kanamycin resistance; (b) the gene for hygromycin resistance plus the insertion of autonomous transposable element Ac within the kanamycin-resistance gene; (c) the gene for hygromycin resistance plus the insertion of the non-autonomous transposable element Ds within the kanamycin-resistance gene. (2) Growth of protoplasts containing the constructs (a), (b) or (c) on media containing either hygromycin only or kanamycin plus hygromycin.

Table 5.3 Transposition of the autonomous maize transposable element Ac (a) and non-autonomous deletion mutants of Ac (b) in heterologous plant species

(a)

Plant species	Assay for transposition[a]	Excision frequency (%)[b]	Integration[c]
Nicotiana tabacum	NPTII	27–70	+
	HPT	36	+
	GUS	33	ND
Arabidopsis thaliana	mol.	30	+
	NPTII	51	+
Solanum tuberosum	NPTII	50	+
Lycopersicon esculentum	mol.	80	+
Glycine max	GUS	45	+
Oryza sativa	HPT	35	+

(b)

Plant species	Activator	Target	Assay for transposition[a]	Excision frequency (%)[b]	Integration[c]
Nicotiana tabacum	Ac	AcΔ(3 kb)	NPTII	50	ND
	Ac-18[d]	AcΔ(3 kb)	NPTII	24–50[e]	+
Arabidopsis thaliana	sAc[f]	Ds	SPT	49[e]	+
Lycopersicon esculenta	Ac	Ds-1	mol.	100	+
	Ac	Ds-202[g]	mol.	ND	+

[a] NPTII, neomycin phosphotransferase; HPT, hygromycin phosphotransferase; GUS, glucuronidase; mol., molecular analysis; SPT, streptomycin phosphotransferase.
[b] Proportions of primary transformants or regenerating calli.
[c] Integration of transposon confirmed by molecular analysis.
[d] Immobilized Ac with a deletion of 4 bp from terminal inverted repeat.
[e] Proportion of progeny in crosses.
[f] Immobilized Ac due to a deletion of 175 bp of 3' end.
[g] Contains bacterial β-glucosidase gene.
ND, not determined.

The results listed in Table 5.3 show that the Ac monocotyledon transposon from maize is able to transpose in a range of dicotyledon species, including tobacco. This opens up the possibility of manipulating the transposon in other species. Table 5.3 also lists a number of examples in tobacco and tomato (*Lycopersicon esculentum*) of mobilization of deleted (non-autonomous) forms of Ac (AcΔ) by another transposase-producing element (Ac). The deleted Ac elements were either produced experimentally (AcΔ) or were the naturally occurring deletion forms (Ds). Two of the Ac elements used for transposase production (Ac-18, sAc) were stabilized by the deletion of one of the 11-bp terminal IRs, which prevents autonomous transposition of Ac-18 and sAc.

Transposon tagging

Since transposable elements can be introduced and mobilized in a plant, screening procedures can be devised to isolate insertional mutations in genes of known function. This strategy for cloning genes is known as transposon tagging and it is dependent upon the gene in question having a known and easily identified mutant phenotype. The technique has been used successfully to clone the *DRL1* locus of *Arabidopsis*, mutations of which have abnormal leaves and roots and no inflorescence.

An outline of a transposon tagging programme is shown in Figure 5.4. A Ds element and a stabilized Ac (sAc) element have been introduced separately into the genome of separate wild-type (normal) plants. These plants are crossed and some of the resulting F_1 progeny will have inherited both the sAc and the Ds element. Since sAc produces transposase, the Ds element is mobilized in these F_1 plants and may by chance be inserted into the gene of interest (*Ph*) to produce an insertional mutation, *Ph*::Ds. If this mutation produces a recessive allele that gives rise to a change of phenotype in homozygous individuals, plants containing the *Ph*::Ds mutation can be identified by crossing the F_1 plants with homozygous stable recessive plants (*ph/ph*).

Most of the progeny of this cross will be heterozygous, *Ph/ph*, with a normal phenotype but a few plants will inherit the *Ph*::Ds allele from the F_1 parent and, since the other parent donates a stable recessive allele (*ph*), these plants (*Ph*::Ds/*ph*) will have a mutant phenotype.

It has been found that the efficiency of this technique can be improved by selecting plants that contain a Ds linked to the gene of interest. This is because transposition occurs into adjacent or flanking regions of DNA more frequently than into other parts of the genome. In *Arabidopsis* the rate of Ds transposition is not very high and it has been possible to increase this by changing the transposase promoter within the Ac element. Thus substituting the cauliflower mosaic virus (CaMV) 35S constitutive promoter and removing all but 65 bp of the Ac transposase 5' UTR improved the frequency of Ds transposition in *Arabidopsis*.

5.4 Retrotransposons

Retrotransposons have been identified in higher plants by homology with other known retrotransposons. Figure 5.5 shows a comparison of the struc-

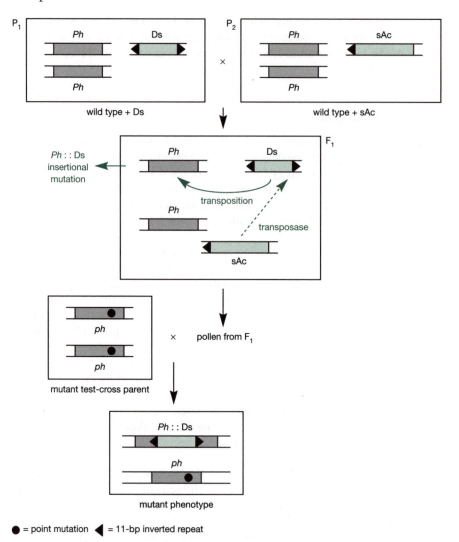

Figure 5.4 Transposon tagging. Schematic diagram of a strategy to use the cloned non-autonomous transposable element Ds to identify clones of the gene *Ph* by marking the gene using insertional mutagenesis. P_1 and P_2, parent plants containing wild-type dominant *Ph* alleles and non-mobile transposable elements (Ds or sAc). F_1, first filial generation containing Ds (with two inverted repeats) and sAc (producing transposase); transposition of Ds in these plants may give rise to *Ph* insertional mutations which can be identified by crossing with a homozygous mutant test-cross plant.

ture of the well-studied retrotransposon, copia, from the fruit fly, *Drosophila melanogaster*, and two putative retrotransposons from plants. The Ta1-3 sequence comes from a family of three members found in the *Arabidopsis*

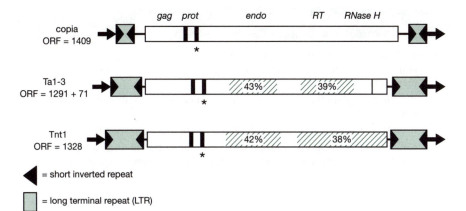

Figure 5.5 Structure of copia-type plant retrotransposons: Ta1-3 from *Arabidopsis* and Tnt1 from tobacco compared with the structure of copia from *Drosophila melanogaster*. The nucleic-binding domain of the *gag* region is shown by a black box and the region forming the protease active site is shown by a black box plus asterisk. The open reading frame (ORF) sizes are given in numbers of amino acids and the regions of highest homology shaded, with the percentage amino acid homology shown. Regions in copia that have homology to retroviral sequences are shown: *gag* encodes components of the virion core; *prot*, *endo*, *RT* and RNaseH encode a protease, endonuclease, reverse transcriptase and ribonuclease respectively. (Modified from Grandbastein, M-A. (1992) Retroelements in higher plants. *Trends in Genetics*, **8**, 103–108.)

genome and Tnt1 comes from the tobacco genome where it occurs in several hundred copies.

Retrotransposons like copia are structurally related to retroviruses and copia is known to transpose via an RNA intermediate. They have long terminal repeats, which are themselves bounded by IRs. They also have sequences with homology to the retrovirus ORFs, which encode proteins with protease, endonuclease, reverse transcriptase and ribonuclease functions. However they lack sequences with homology to the *env* region of a retrovirus. This region is responsible for the envelope of the retrovirus and its absence from retrotransposons means that they are unlikely to generate virus-like particles. The similarity between retrotransposon and retrovirus structure suggests an evolutionary relationship.

Figure 5.5 shows the structural organization of copia together with the regions of homology found in Ta1-3 and Tnt1. Both Ta1-3 and Tnt1 have the nucleic acid-binding region of the retroviral *gag* gene as well as the conserved region forming the active site of the retroviral protease. They also have regions of homology with the deduced amino acid sequences of the endonuclease, reverse transcriptase and ribonuclease retroviral regions of copia. Tnt1 is the only plant retrotransposon known to be transcriptionally active but transposition of the plant sequences has not been recorded.

Major Learning Objectives for Chapter 5

1. Understand the structure of the Ac and Ds transposable elements in maize and comprehend the role of different regions of the Ac element in the mechanism of transposition.

2. Be able to explain the role of transposons in causing unstable insertional mutations.

3. Be able to distinguish between different types of plant transposable elements.

4. Understand the technique of transposon tagging for the identification and isolation of genomic clones containing genes with a known mutant phenotype.

Further reading

BALCELLS, L., SWINBURN, J. and COUPLAND, G. (1991). Transposons as tools for the isolation of plant genes. *Trends in Biotechnology*, **9**, 31–37.
A well written description of this technique containing more detail than Grierl and Saedler (1992).

BANCROFT, I., JONES, J.D.G. and DEAN, C. (1993). Heterologous transposon tagging of the *DRL2* locus in *Arabidopsis*. *Plant Cell*, **5**, 631–638.
Describes the use of transposon tagging to clone the *Arabidopsis* locus (gene) *DRL2*, mutations of which have an altered developmental pattern.

GIERL, A. and SAEDLER, A. (1992). Plant-transposable elements and gene tagging. *Plant Molecular Biology*, **19**, 39–49.
A good general review covering the material of this chapter.

GRANDBASTIEN, M-A. (1992). Retroelements in higher plants. *Trends in Genetics*, **8**, 103–108.
A good account of this type of transposable element, which is not covered by the other references.

HEHL, R. (1994). Transposon tagging in heterologous host plants. *Trends in Genetics*, **10**, 385–386.
An explanation of the modifications of Ac, which improve transposition of transposable elements in the heterologous species *Arabidopsis*.

Part two

Agrobacterium tumefaciens

Chapter 6

Agrobacterium tumefaciens

6.1 Crown gall disease

Agrobacterium tumefaciens is a free-living Gram-negative soil bacterium. Virulent strains of this bacterium are able to infect wounded dicotyledon plants and induce the production of a neoplastic growth, or gall, by the plant. Since plants commonly suffer small abrasions at soil level due to wind movement, these galls often grow at soil level where the *Agrobacterium* infects the plant wound, and the disease is consequently known as crown gall disease. *A. tumefaciens* can infect a wide range of dicotyledon species but is not a disease of monocotyledons.

Cells from the *Agrobacterium*-induced plant gall, unlike the cells from normal healthy plant tissue, have the ability to proliferate and grow in culture without the addition of the plant growth hormones auxin and cytokinin.

Virulent *Agrobacterium* strains contain a very large plasmid, known as the the tumour-inducing or Ti plasmid. These plasmids vary in size between 130 and 230 kb in individual strains of bacteria. Non-virulent *Agrobacterium* strains, which are unable to infect plants, contain no Ti plasmid.

The properties of the plant gall, such as morphology (rough or smooth) and the ability to synthesize unusual amino acids, known collectively as opines, are also determined by the Ti plasmid. The structure of the opines nopaline and octopine (so named because it has also been isolated from the octopus) are shown in Figure 6.1.

A strain of *Agrobacterium* that causes the growth of a gall producing a particular opine (e.g. octopine) will also be able to catabolize this opine and this property is also encoded on the Ti plasmid. In other words an *Agrobacterium* strain harbouring an octopine Ti plasmid will cause the formation of octopine-synthesizing galls and the bacteria will then be able to utilize this unusual amino acid. In contrast an *Agrobacterium* strain with a nopaline Ti plasmid results in the formation of nopaline-synthesizing galls and of bacteria capable of utilizing nopaline. Since the plant host is unable to catabolize opines, the invading bacteria have effectively hijacked the plant's metabolism to produce a compound which only they can utilize.

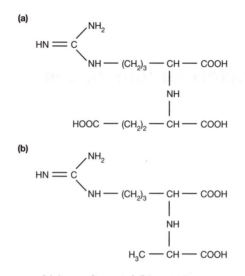

Figure 6.1 Structure of (a) nopaline and (b) octopine.

If the cells from a gall are grown in culture in the presence of an antibiotic (such as cefotaxime), which prevents the proliferation of the bacteria but not the plant cells, it is possible to produce bacteria-free crown gall cell cultures. It has been shown that these uninfected plant cells contain a region of the Ti plasmid that has become integrated into the plant nuclear genome. This region of the Ti plasmid is known as the transfer or T-DNA. Plant cells such as these, which have genetic information from another organism integrated into their chromosomes, are said to be transformed.

Box 6.1 Experiment to demonstrate that crown gall cells contain T-DNA

An experiment demonstrating the presence of T-DNA in a plant genome is shown in Figure a. DNA was extracted from *Agrobacterium* containing a Ti plasmid, from crown gall cells that had been infected by this strain of *Agrobacterium* and then cured, and from healthy non-infected plants. Following digestion of these DNA samples by a restriction endonuclease and size fractionation of the resulting DNA fragments, the DNA was transferred to a membrane by Southern blotting and probed with T-DNA. The autoradiogram illustrated in Figure a shows the T-DNA fragments from the Ti plasmid in *Agrobacterium* (lanes 1 and 2) and T-DNA fragments in the infected plant genomic DNA (lanes 3, 5 and 6), with non-infected (control) plant DNA showing no hybridization to the T-DNA probe (lane 4).

▶

(Box 6.1 continued)

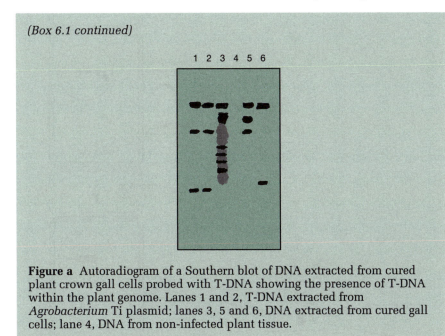

Figure a Autoradiogram of a Southern blot of DNA extracted from cured
plant crown gall cells probed with T-DNA showing the presence of T-DNA
within the plant genome. Lanes 1 and 2, T-DNA extracted from
Agrobacterium Ti plasmid; lanes 3, 5 and 6, DNA extracted from cured gall
cells; lane 4, DNA from non-infected plant tissue.

A summary of *Agrobacterium* infection and the production of hormone-
independent crown gall tissue is shown in Figure 6.2.

6.2 Ti plasmid genetic structure

A comparative analysis of the structure of the octopine Ti plasmid, pTiAch5, and
the nopaline Ti plasmid, pTiC58, is shown in Figure 6.3. These two plasmids
differ in size (pTiAch5 is about 180 kb and pTiC58 is about 200 kb) but there are
regions of homology between the two where they share common DNA sequences
and these regions are shown in grey in Figure 6.3. A number of insertion mutations of these Ti plasmids have been produced using a bacterial transposon. The
sites of those mutations leading to the loss of oncogenicity (ability to form a gall)
are arrowed in Figure 6.3 and it can be seen that all these mutations fall within
the areas of homology. The T-DNA regions of these Ti plasmids also contain
homologous sequences and mutation sites that lead to loss of oncogenicity
(Figure 6.3). The genes for opine catabolism, as expected, share no homology in
these two plasmids.

A consensus genetic map of the octopine Ti plasmid is shown in Figure 6.4
with the functionally important regions labelled. In common with other plasmids, Ti has a region specifying conjugal transfer function and an origin of
replication. Four other regions are marked: the bipartite T-DNA region (T_L
plus T_R), an enhancer sequence that increases T-DNA transfer, the position of
the gene or genes for opine catabolism and the virulence (*vir*) region.

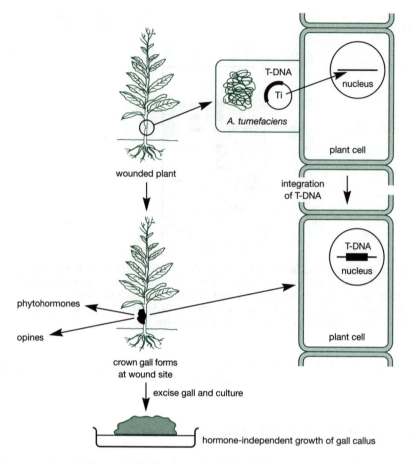

Figure 6.2 Summary of *Agrobacterium* infection and the production of hormone-independent crown gall tissue.

T-DNA structure and function

The octopine Ti plasmid, pTiAch5, contains two T-DNA regions (shown in Figure 6.5) called T_L and T_R but only the T_L DNA region is oncogenic. The transfer of these regions into plant nuclear genomic DNA is independent. Both T_L and T_R are bounded by a 24-bp direct repeat. These are not perfect repeat sequences but they have sequence homology with the equivalent T-DNA direct repeat border sequences of the nopaline Ti plasmid (Figure 6.6). The T_L octopine T-DNA region contains eight ORFs (Figure 6.5), which are transcribed when the T-DNA is integrated into the plant nuclear genome. These ORFs have features of eukaryotic plant genes rather than prokaryote (bacterial) genes; for example, they contain polyadenylation signals. The function of some of these ORFs (genes) is known:

$\left.\begin{array}{l} auxA \\ auxB \end{array}\right\}$ phytohormone (auxin) biosynthesis

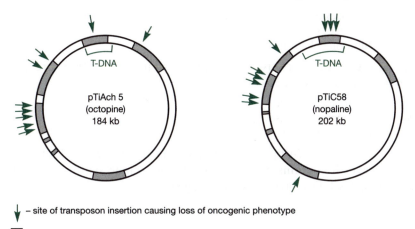

– site of transposon insertion causing loss of oncogenic phenotype

– regions of homology between plasmids

Figure 6.3 Comparison of the structural organization of the octopine Ti plasmid pTiAch5 and the nopaline Ti plasmid pTiC58. Areas of DNA homology shown as solid blocks; positions where insertional (transposon) mutations lead to loss of oncogenicity shown with vertical arrows; T-DNA regions shown with an open block.

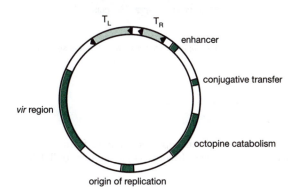

Figure 6.4 Genetic map of an octopine Ti plasmid.

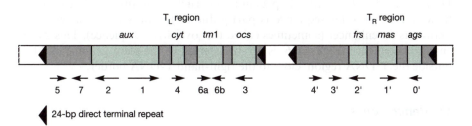

Figure 6.5 Genetic map of the T-DNA regions of an octopine Ti plasmid. Open reading frames shown as arrows 1–7 in T_L region and 0'–4' in T_R region. Regions of known function shown as green blocks: *aux*, auxin biosynthesis; *cyt*, cytokinin biosynthesis; *tm*1, tumour gall size; *ocs*, octopine synthase; *frs*, fructopine synthase; *mas*, mannopine synthase; *ags*, agropine synthase.

cyt	phytohormone (cytokinin) biosynthesis
ocs	octopine synthase
*tm*1	function unknown, mutations affect gall size

In the T_R region there are five ORFs and it is known that three are genes responsible for the biosynthesis of further opines:

frs	fructopine synthase
mas	mannopine synthase
ags	agropine synthase

The T_R region has the 24-bp direct repeats and can be integrated into plant DNA; however it contains no genes for phytohormone synthesis and this explains why this T-DNA region is not oncogenic.

(a)

nopaline	TGGCAGGATATATTGTGGTGTAAAC	left
	TGACAGGATATAT–GGCGGGTAAAC	right
octopine	CGGCAGGATATAT–CAATTGTAAAT	left
	TGGCAGGATATAA–CCGTTGTAATT	right

Bold bases show differences from the nopaline left border direct repeat sequence

(b)

enhancer sequence	TAAGTCGCTGTGTATGTTTGTTTG

Figure 6.6 (a) T-DNA border direct repeat sequences from a nopaline and an octopine Ti plasmid. (b) Enhancer (overdrive) sequence for T-DNA transfer.

It is known that the right-hand T-DNA border repeat is essential for T-DNA transfer but the left-hand border repeat is not vital. The transfer of DNA from *Agrobacterium* Ti plasmid constructs containing only the right-hand T-DNA border sequence is partly due to the presence of a sequence known as an enhancer (sometimes called the 'overdrive' sequence). This 24-bp enhancer sequence (Figure 6.6) is not transferred to the plant genome and lies outside the T-DNA region, close to the right-hand T-DNA border.

Virulence genes

The Ti plasmid genes responsible for the transfer of T-DNA from the *Agrobacterium* plasmid to genomic DNA within the nucleus of a plant cell are situated within a single 40-kb region; 24 virulence (*vir*) genes in eight operons have been identified (Figure 6.7).

The function of some of the *vir* genes is known:

*vir*A ⎱
*vir*G ⎰   two-component regulatory system

*vir*D1 ⎱
*vir*D3 ⎰   cooperate with *vir*D2 to produce a DNA-binding complex that produces single-strand nicks in the DNA duplex

*vir*D2   covalently bound to 5' end of single-stranded (ss)DNA

*vir*C1   binds to the T-DNA transfer enhancer sequence

*vir*E2   ssDNA-binding protein

*vir*B operon   includes genes encoding predicted membrane proteins and possibly involved in producing a structure that allows the transfer of DNA from bacterium to the plant cell

*vir*B11   ATPase

*vir*F   involved in host range specificity

Similarities between the Ti *vir* genes and the genes involved in the conjugative transfer of DNA between bacteria have been noticed, particularly with the *Tra* genes of the broad host range plasmid, RP4, and these similarities have helped to formulate a model for the mechanism of T-DNA transfer (see section 6.4).

6.3 Plant wound signals and control of *vir* gene expression

Wounded plant cells release phenolic compounds, the structure of one of which (acetosyringone) is shown in Figure 6.8. These phenolic compounds will attract *Agrobacterium* to the wound site and are involved in the induction of Ti *vir* genes. The *vir*A and the *vir*G genes, which form the two-component control complex, are expressed during free-living growth of *Agrobacterium* but the other *vir* genes are induced by the wound phenolic compounds. The *vir*A gene

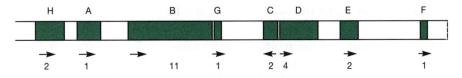

Figure 6.7 Map of the *vir* gene cluster in the octopine Ti plasmid. Letters show each operon, arrows show the direction of transcription and the numbers show the number of genes in each operon.

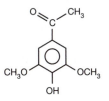

Figure 6.8 Structure of acetosyringone.

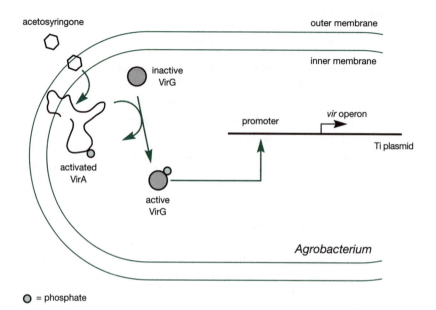

Figure 6.9 Activation of *vir* genes via VirA and VirG.

product (VirA) is present in the bacterial inner membrane and acetosyringone causes the autophosphorylation of this protein. The phosphorylated form of VirA can in turn phosphorylate the *virG* gene product (VirG), which thus becomes activated. The active VirG protein is a positive transcription factor involved in the activation of transcription (expression) of all the other *vir* genes. Figure 6.9 shows this signal transduction pathway.

6.4   Molecular mechanism of T-DNA transfer

Our understanding of the molecular mechanism of T-DNA transfer from *Agrobacterium* to plant nuclear chromosomes is far from complete but some of the components of the process can be identified and a model of part of the mechanism that has wide acceptance is outlined in Figure 6.10. In this model single-stranded nicks are made in the T-DNA border repeat by the VirD1, VirD2, VirD3 protein complex (Figure 6.10a). The VirD2 protein then becomes covalently linked to the 5' end of the displaced ssDNA molecule (Figure 6.10b).

As the ssDNA–VirD2 molecule is displaced from the DNA duplex, DNA repair synthesis replaces this DNA strand and the displaced strand becomes bound to VirE2 proteins, which presumably stabilize the single-stranded molecule (Figure 6.10c). It is known that in *Agrobacterium* with active *vir* genes, both single-stranded and double-stranded copies of the T-DNA

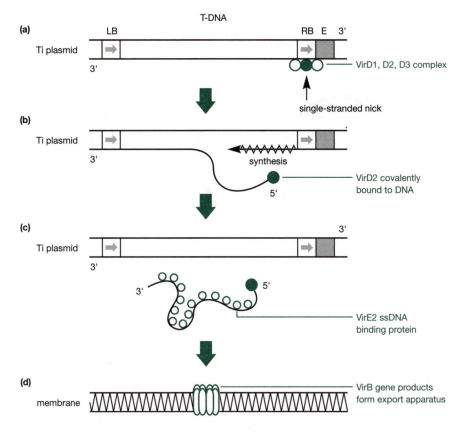

Figure 6.10 Model for the formation of single-stranded T-DNA in *Agrobacterium* showing the putative function of VirD1, VirD2, VirD3, VirE2 and the *vir*B operon. See text for explanation of stages (a) – (d).

region exist. A membrane structure responsible for the export of T-DNA from the bacterium is constructed from the proteins encoded by the *vir*B operon (Figure 6.10d).

Another class of genes involved in the infection and transformation process have been identified in *Agrobacterium*. The genes, *chv*A and *chv*B, are not present on the Ti plasmid but are situated on the bacterial chromosome. These two genes are necessary for attachment of *Agrobacterium* cells to the plant cell wall. The gene *chv*B encodes a protein involved in cyclic β-1,2-glucan synthesis, while *chv*A is involved in the transport of this compound to the bacterial periplasmic space. However the precise role of this glucan compound in cell adhesion and infection is not known.

The process of *Agrobacterium* infection and plant cell transformation outlined in this chapter is summarized in Figure 6.11.

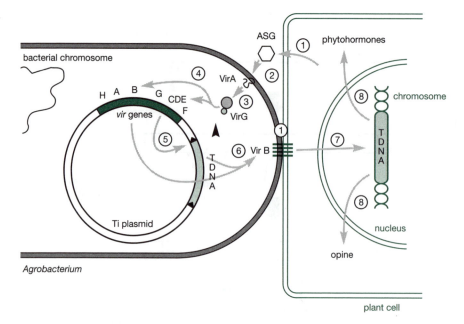

Figure 6.11 Outline of the process of *Agrobacterium* infection and plant cell transformation. The numbers refer to sequential stages: 1, production of acetosyringone (ASG) by the plant and attachment of *Agrobacterium* to plant cell wall; 2, VirA present on the *Agrobacterium* inner membrane is autophosphorylated in response to acetosyringone produced by the plant; 3, VirG is phosphorylated by VirA; 4, phosphorylated VirG activates the transcription of other *vir* genes; 5, VirD and VirE proteins produce single-stranded T-DNA; 6, VirB proteins produce the export apparatus; 7, T-DNA becomes integrated into the plant nuclear DNA; 8, plant phytohormones and opines are produced by T-DNA genes;

Major Learning Objectives for Chapter 6

1. Knowledge of the infection of dicotyledon plants by *Agrobacterium tumefaciens* and the symptoms of crown gall disease.

2. Knowledge of the overall structure of the Ti plasmid.

3. Detailed knowledge of the structure of T-DNA and understand the role of T-DNA genes in determining oncogenicity.

4. Be able to explain:
    (a) the function of the Ti plasmid virulence genes, and
    (b) the control of their expression by plant phenolic compounds.

5. Be able to use the information in 1–4 to produce an explanation of plant cell transformation by virulent strains of *Agrobacterium*.

Further reading

HOOYKAS, P.J.J and SCHIPEROOT, R.A. (1992). *Agrobacterium* and plant genetic
engineering. *Plant Molecular Biology*, **19**, 15–38.
A comprehensive account of *Agrobacterium* Ti plasmid plant transformation
covering most of the material of this chapter.

HOOYKAS, P.J.J. and BEIJERSBERGEN, A.G.M. (1994). The virulence system of
Agrobacterium tumefaciens. *Annual Review of Phytopathology*, **32**, 157–180.

LESSL, M. and LANKA, E. (1994). Common mechanisms in bacterial conjugation and
Ti-mediated T-DNA transfer to plant cells. *Cell*, **77**, 321–324.
These last two references give a clear description of our knowledge of the
mechanism of plant cell transformation by the *Agrobacterium* Ti plasmid.

Chapter 7

Ti plasmid as a vector for plant transformation

7.1 Plant transformation

The natural mechanism of *Agrobacterium* Ti plasmid transformation of plant cells has been manipulated by recombinant DNA technology to produce a number of vector systems allowing the introduction of foreign DNA into higher plant nuclear genomes. A variety of vector systems exist but in this chapter the widely used binary vector system based on the BIN19 plasmid will be described.

The technique of introducing foreign DNA into plant cells (transformation) is an extremely powerful tool for the investigation of plant genes, as well as an important method for plant biotechnology. Examples of the use of this technique will occur throughout the rest of this book but in this chapter the use of transformation techniques in investigative science will be covered in order to explain the broad range of different experimental uses of plant cell transformation. The production of transgenic plants by Ti plasmid transformation for crop improvement or biotechnology will be covered in Chapters 15 and 16.

7.2 A binary vector system

The development of a binary (two part) vector system from the *Agrobacterium* crown gall disease was based on two features of the mechanism of transfer of T-DNA from the Ti plasmid to plant nuclear chromosomes.

1. The only T-DNA sequences necessary for T-DNA integration into the plant chromosomal DNA are the terminal repeat sequences. This means that the Ti plasmid genes between the T-DNA border repeats can be replaced by other genetic information and these genes will consequently be transferred to plant chromosomes.
2. The virulence (*vir*) genes, which produce the machinery for T-DNA transfer, act in *trans*, that is these genes do not have to be situated on the same plasmid as the T-DNA border repeats.

The necessary characteristics of a vector system for plant transformation based on the *Agrobacterium* Ti plasmid can be listed.

1. Removal of oncogenic genes *aux*A, *aux*B and *cyt* from the T-DNA. This prevents the over-production of phytohormones, which cause the growth of undifferentiated cells. Phytohormone concentrations and ratios can then be manipulated *in vitro* to induce the growth of differentiated shoots and roots (section 7.3).
2. Presence of a multiple cloning site (MCS), consisting of a group of unique restriction enzyme cutting sites, within the T-DNA border repeats, to allow the insertion of novel DNA for transformation.
3. Presence of a selectable marker within the T-DNA border repeats that allows the selection of plant cells which have been transformed with the T-DNA construct. The selectable markers are often antibiotic-resistance genes (kanamycin resistance, hygromycin resistance) of bacterial origin. These genes have to be 'converted' into plant genes by the addition of appropriate plant signals (a plant promoter and a polyadenylation signal) for gene expression.
4. The T-DNA construct detailed in (1)–(3) should be on a small plasmid that can be easily manipulated in *E. coli*. This plasmid should contain the usual features of a plasmid cloning vector, namely high copy number replication, a selectable marker (antibiotic resistance) for bacterial transformation and limited restriction endonuclease sites, so that the restriction sites in the MCS are unique.
5. This small plasmid containing the T-DNA construct must be able to replicate in *E. coli*, to allow manipulation of the DNA, and in *Agrobacterium*, which will transfer the T-DNA construct to the plant cell.
6. The *Agrobacterium* strain used must contain a disarmed Ti plasmid; this plasmid contains all of the virulence (*vir*) genes but has no T-DNA. It is therefore not oncogenic but is capable of mobilizing the T-DNA construct present in the other smaller and more easily manipulated plasmid.

An example of a pair of binary vector plasmids is shown in Figure 7.1. The small plasmid pBAG3 (13.5 kb), containing the T-DNA border repeats (LB and RB), is one of a generation of plasmids all derived from pBIN19. Between the T-DNA left-hand border (LB) and right-hand border (RB) repeats it contains the following.

1. An MCS containing 21 unique restriction endonuclease cutting sites,
2. This MCS is within the *lacZα* sequence; this sequence encodes the N-terminal portion of *E. coli* β-galactosidase, which can complement the appropriate *E. coli* host to produce active β-galactosidase in the bacterial transformants. β-Galactosidase activity can be detected in bacterial colonies by the production of a blue precipitate and this will be a property only of bacteria containing non-recombinant plasmids. Plasmids that have DNA inserted into the MCS will effectively have an insertional mutation of *lacZα* and these will not complement the *E. coli* host β-galactosidase mutant gene. *E. coli* transformed with recombinant

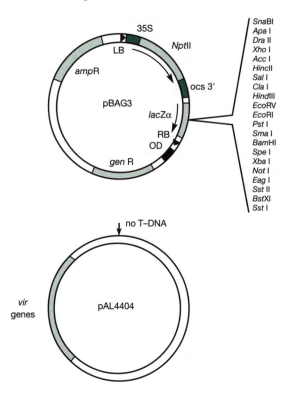

Figure 7.1 Binary vectors for plant transformation. pBAG3 plasmid containing left-hand border (LB) and right-hand border (RB) T-DNA repeats and a multiple cloning site (MCS). pBAG3 can replicate in *Escherichia coli* and *Agrobacterium*. pAL4404, disarmed helper plasmid in *Agrobacterium. amp*R, ampicillin resistance; 35S, constitutive promoter from cauliflower mosaic virus; *Npt*II, neomycin phosphotransferase gene; *ocs* 3', 3' non-coding region from octopine synthase gene; OD, overdrive sequence; *lacZα*, 5' coding sequence plus promoter of *lacZ* gene; *gen*R, gentamicin resistance; *amp*R, ampicillin resistance; *vir*, virulence genes.

plasmids will therefore produce white colonies. This blue/white test provides selection for recombinant pBAG3 plasmids.

3. The coding sequence of *Npt*II, the bacterial neomycin phosphotransferase gene, which confers kanamycin resistance to cells. In order that this gene is expressed in plant cells it has the 35S promoter from cauliflower mosaic virus (CaMV) at its 5' end. This is a promoter that confers a high level of gene expression in all parts of the plant, is not influenced by environmental factors and is therefore called a constitutive promoter. Although the promoter comes from a virus, it is a plant-like promoter which in its natural viral genome allows the viral proteins to be produced by the plant cell transcription and translation machinery. The *Npt*II gene also has the 3' non-coding region (including a polyadenylation signal) from a Ti plasmid octopine synthase gene (*ocs*) inserted at the 3' end of the coding sequence. This sequence ensures the correct termination of transcription and addition

of a poly(A) tail to the *Npt*II mRNA. This *Npt*II chimeric gene will be expressed and confer kanamycin resistance on plant cells.

Outside the LB and RB T-DNA sequences, pBAG3 contains the enhancer or overdrive (OD) sequence near the RB as well as two bacterial antibiotic genes (*amp*R, ampicillin resistance; *gen*R, gentamicin resistance) as alternatives for selection of the plasmid in *E. coli* and *Agrobacterium*.

The disarmed Ti plasmid shown in Figure 7.1 (pAL4404) is a derivative of pTiAch5 that has the T-DNA region deleted but which has all the functional *vir* genes.

Two other components are needed for a Ti plasmid-based plant transformation system. One of these is a method of transferring the small plasmid containing the T-DNA construct from *E. coli* to *Agrobacterium*, and the most commonly used method for this is called electroporation. The other component needed is discussed in the next section.

Box 7.1 Electroporation

A culture of *Agrobacterium* is suspended in a solution of plasmid DNA and a pulse of high-voltage electric current passed through the suspension. The electric current punches small holes in the bacterial cell walls and the bacteria are therefore capable of taking up plasmid DNA from the solution and subsequently repairing the cell wall damage. The transformed bacteria are plated out on to Petri dishes containing an antibiotic, where only bacteria containing the plasmid with its antibiotic-resistance gene are capable of growing and producing colonies.

Figure 7.2 gives an outline of the important steps in a protocol to produce an *Agrobacterium* culture that contains a binary vector for transforming plant cells.

7.3 Plant transformation: manipulation of plant cells and explants

The other important component of a Ti plasmid-based plant transformation system is a method of manipulating and culturing plant cells and tissues. A method of infecting wounded tissue with *Agrobacterium* is required first and this must be followed by a method of selecting the plant cells that have been transformed from those that have not. The method of selection of plant cell transformants is commonly growth on an antibiotic-containing medium, where resistance to the antibiotic is conferred by the T-DNA construct. The growth of plant cells is not affected by all antibiotics and therefore antibiotics like kanamycin and hygromycin have been chosen because they are known to be inhibitory to plant growth. Lastly the transformed plant cells must be cultured so that they will grow and differentiate to produce a normal, fertile plant.

Although *A. tumefaciens* will infect a very wide range of dicotyledon species not all plants are amenable to tissue culture. Members of the Solanaceae, such

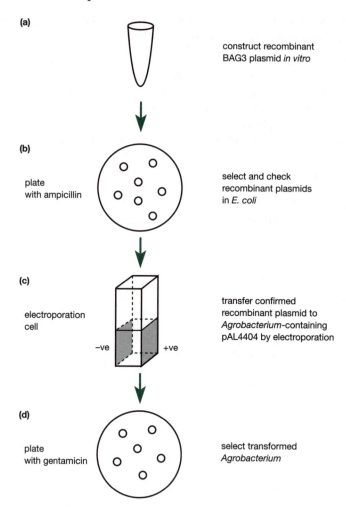

(a) construct recombinant BAG3 plasmid *in vitro*

(b) plate with ampicillin — select and check recombinant plasmids in *E. coli*

(c) electroporation cell — transfer confirmed recombinant plasmid to *Agrobacterium*-containing pAL4404 by electroporation

−ve   +ve

(d) plate with gentamicin — select transformed *Agrobacterium*

Figure 7.2 Outline of the production of an *Agrobacterium* culture that contains a binary vector for plant transformation.

as tobacco, potato, petunia and tomato, are particularly easy to manipulate in culture and a great deal of the plant transformation experimentation has been done with these species, particularly tobacco.

Figure 7.3 shows a protocol for the transformation of tobacco leaf cells and the subsequent regeneration of normal flowering plants from the transformed cells. These plants, which contain foreign DNA, are known as transgenic plants and the introduced gene is often referred to as the transgene.

Small pieces of leaf tissue are cut from a tobacco leaf using a cork-borer (Figure 7.3a) and these are incubated in a suspension of *Agrobacterium* containing both recombinant pBAG3 and the helper Ti plasmid pAL4404 (Figure 7.3b). The bacteria are attracted to cells at the wounded tissue surface and the T-DNA construct is transferred to the plant cell nuclear chromosomes (see Chapter 6).

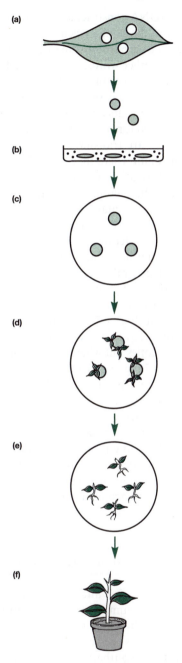

Figure 7.3 Protocol for the transformation of tobacco leaf cells using an *Agrobacterium* binary vector system and the selection and regeneration of transgenic flowering plants. (a) Tobacco leaf; (b) *Agrobacterium* + leaf discs; (c) tissue culture medium; (d) tissue culture medium + kanamycin + cefotaxime + auxin (0.03 mg l⁻¹) + cytokinin (1.0 mg l⁻¹); (e) tissue culture medium + kanamycin + cefotaxime + auxin (3.0 mg l⁻¹) + cytokinin) 0.02 mg l⁻¹); (f) transgenic tobacco plant.

The site of insertion of the T-DNA construct appears to be largely random and multiple insertions may occur in the same nucleus. Sometimes the T-DNA construct will insert into a region of a chromosome that is heterochromatic and contains no functional genes. In other words, it is possible for the T-DNA to go into a chromosomal region that does not provide the correct chromosomal architecture (chromatin structure) for gene expression. Clearly, if such transformed cells are produced, the antibiotic-resistance selection gene in the construct will not be expressed and such cells will therefore not be recovered (selected).

After infection the leaf pieces are transferred to a simple nutrient salt and sucrose plant tissue culture medium where they can be left for a short period to ensure that infection and transformation have occurred (Figure 7.3c). From this medium they are put on to a medium containing kanamycin to select for transformed cells. The medium must also contain an antibiotic, such as cefotaxime, to eliminate *Agrobacterium*. This antibiotic must not affect the growth of the plant cells, which must be allowed to proliferate. At this stage the plant tissue culture medium is supplemented with plant hormones, which will stimulate the development of differentiated shoots. In general, plant development can be controlled by the ratio of the concentration of the phytohormone auxin to the concentration of the phytohormone cytokinin. Thus shoot development is stimulated by about 1 mg l^{-1} cytokinin and 0.03 mg l^{-1} auxin (Figure 7.3d). After shoots have formed at the cut leaf disc rim, they are transferred to a new medium, which still contains the kanamycin and cefotaxime antibiotics but where the cytokinin and auxin concentrations have been changed to 0.02 mg l^{-1} and 3.0 mg l^{-1} respectively (Figure 7.3e). The small rooted transgenic plantlets that are formed can finally be transferred to a soil compost for normal growth (Figure 7.3f).

This procedure has been modified for the transformation of other species and the details will depend primarily on the facility with which the particular species can grow and differentiate in culture. It has recently been shown that transgenic *Arabidopsis* plants can be produced by simply vacuum infiltrating the developing flowers of pot (soil) grown plants and screening the seeds for the ability to germinate in the presence of the antibiotic (NB *Arabidopsis* is a species that can self-fertilize).

7.4   Optimization of gene expression in transgenic plants

The protocol described in sections 7.3 and 7.4 produces plants capable of growth and development on a medium containing an antibiotic that will kill the non-transgenic parent plants; however the presence of T-DNA construct and expression of the transgenes must be verified. Figure 7.4 shows an analysis of 17 transgenic potato plants. A Southern blot of restriction endonuclease-digested genomic (chromosomal) DNA, extracted from each of the 17 transgenic potato plants, has been probed with the RB of T-DNA so that the number of bands in the autoradiogram corresponds to the number of copies of the T-DNA construct present in each transgenic plant. It can be seen that the plant in lane 1 has three bands, corresponding to three T-DNA inserts,

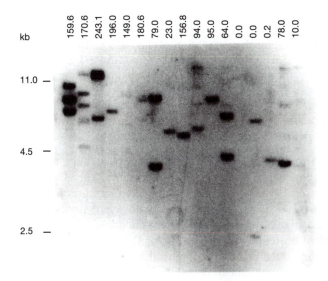

Figure 7.4 Southern blot analysis of transgenic potato plants (cv. Maris Bard) showing the presence of patatin promoter-β-glucuronidase (GUS) construct. The figures above each lane give the levels of glucuronidase enzyme activity in extracts from each plant in nmoles per minute per gram fresh weight. (From Mitten, D.H. Hoon, M., Burrell, M.M. and Blundy, K.S. (1990) Strategies for potato transformation and regeneration. In Vayda, M.E. and Park, W.D. (eds) *The molecular and cellular biology of the potato*, pp. 181–191. CAB International, Wallingford.)

whereas the plant in lane 4 has only one band and therefore only one T-DNA insert. The numbers at the top of the figure show the level of reporter trans-gene expression, given in units of enzyme activity (see Box 7.2 and section 7.5).

This analysis shows that levels of enzyme activity vary from 0 to 243.1 units. There is not an exact correspondence between the number of T-DNA inserts and the level of transgene expression; thus the transgenic plant with one T-DNA insert in lane 4 has 196.0 units of activity, whereas the plant in lane 7, which has two T-DNA inserts, has only 79.0 units of activity. Since all of these plants contain the same T-DNA construct the position of the T-DNA insert within the genome clearly has a major effect on the expression of the genes it contains.

Box 7.2 T-DNA construct of Figure 7.4

The transgene introduced into the potato plants analysed in Figure 7.4 is a chimeric gene consisting of the coding sequence of a bacterial gene for β-glucuronidase (GUS). This enzyme is not found in plants and its production in transgenic plants must therefore be dependent on the expression of the transgene. This GUS coding sequence is commonly used as a reporter of transgene expression since GUS enzyme activity is easily measured. The

(Box 7.2 continued)

promoter controlling the expression of the GUS coding sequence in Figure 7.4 is a potato promoter from a gene encoding a tuber-specific protein (patatin). The 3' UTR comes from a Ti plasmid nopaline synthase gene.

Effect of coding sequence on the expression of non-plant genes in transgenic plants

Poor expression in plants is a frequently reported characteristic of heterologous genes (particularly from non-plant species). The effect of changing the DNA sequence of a gene (*cry*1A(b)) isolated from a Gram-positive bacterium (*Bacillus thuringiensis* var. *kurstaki*) on plant expression was tested in transgenic tomato plants. The gene *cry*1A(b) encodes a protein toxic to insects (insect control protein). Figure 7.5a shows the changes made to this bacterial gene, which included removing potential plant polyadenylation signals and ATTTA sequences, which can destabilize mRNA, removing A/T-rich regions that resemble plant introns, and changing the codons that are rarely used in plants. The changes were made in two stages to produce a partially modified (PM) and fully modified (FM) plant gene. Figure 7.5b shows the effects that these changes have on gene expression in transgenic tomatoes. It can be seen from the histogram that only the modified sequences produce insect control protein above the 10 ng/50 μg total protein level.

Effect of chromatin structure on transgene expression

Attempts have been made to insulate the transgene from the higher order (chromatin) structure of the chromosomal DNA into which the T-DNA is integrated. To do this workers have included scaffold-attachment regions (SARs) flanking the transgene in the T-DNA construct. The argument for this approach is that these sequences are thought to attach the 30-nm chromatin fibre to a proteinaceous scaffold and that the looped domains so formed represent expressed genes (see section 1.3). Including SARs in the construct may produce a new loop domain and remove the variation in expression associated with the position of integration (see section 7.3). However, although this strategy has had some success in animal cell transformation the results in plants are not conclusive.

7.5 The use of transgenic plants to study plant genes

Studies using transgenic plants are now an integral part of plant molecular genetics. We have already seen the use of transformation technology in Chapter 5, where heterologous transposons have been introduced into plants for transposon tagging.

(a)

	Sequence		
	Wild-type gene	Partially modified gene	Fully modified gene
No. of codons different from wild type	—	9.5%	60%
G + C content	37%	41%	49%
No. of plant polyadenylation signal sequences	18	7	1
No. of ATTTA sequences	13	7	0

(b)

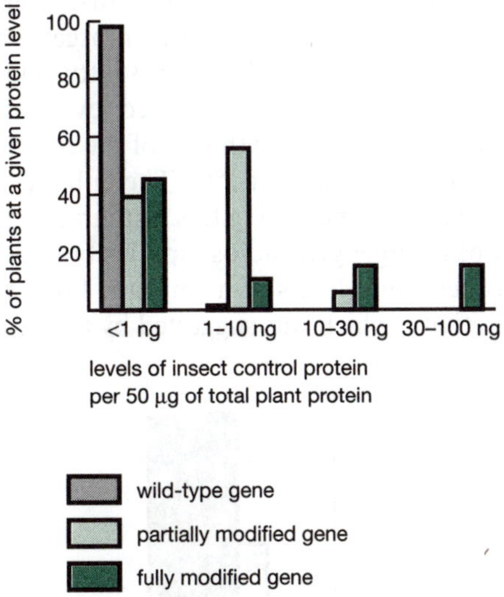

Figure 7.5 (a) Summary of changes made to *cry*1A(b) coding sequence. (b) Distribution of transgenic tomato plants that produce the insect control protein encoded by *cry*1A(b): levels of insect control protein produced in plants containing the wild-type gene, a partially modified gene or a fully modified gene (see Table 7.1). (After Perlack F.J., Fuchs, R.L., Dean, D.A., McPherson, S.L., and Fischoff, D.A. (1991) Modification of the coding sequence enhances expression of insect control protein genes. *Proceedings of the National Academy of Sciences USA*, **88**, 3324–3328.)

In this section further examples of the use of transgenic plants will be described where they are involved in studies of plant gene function and expression. The constructs described can be classified as: (i) addition of a new coding sequence, (ii) addition of a new control sequence, (iii) the use of reporter gene constructs, and (iv) removal of gene function.

Addition of new coding sequence (function)

The enzyme ω-3 fatty acid desaturase converts dienoic fatty acids (16:2 and 18:2) to trienoic fatty acids (16:3 and 18:3) and a cDNA clone for this enzyme has been isolated from *Arabidopsis thaliana*. The trienoic fatty acids (16:3 and 18:3) have higher levels of unsaturation than the dienoic fatty acids (16:2 and 18:2) and the ω-3 fatty acid desaturase therefore increases the level of unsaturation in the fatty acids of chloroplast membranes. A correlation between chilling sensitivity (plant damage by low positive temperatures) and the level of unsaturation of fatty acids in chloroplast membranes has been reported in higher plants.

In order to test the hypothesis that increased levels of fatty acid desaturation give rise to chill tolerance, the ω-3 fatty acid desaturase from a chill-tolerant species (*Arabidopsis*) was used to transform a chill-sensitive species (tobacco). Levels of 16:2, 18:2, 16:3 and 18:3 fatty acids were measured in transgenic plants and compared with the levels in both the wild-type (WT) parent (non-transgenic plants) and plants that had been transformed with the 'empty' vector (EV) (Table 7.1). Table 7.1 shows that the transgenic plants (SRT-1, SRT-5 and SRT-6) had elevated levels of 16:3 and 18:3 fatty acids compared with the controls (WT and EV) and concomitant reduced levels of 16:2 and 18:2 fatty acids. Figure 7.6 shows the relative growth rate of the second leaf of these plants at 25°C, following a 7-day exposure at 1°C. The enhanced growth of the ω-3 desaturase transgenic plants can be seen in this histogram.

This manipulation of a single gene across species boundaries provides strong support for the hypothesis that levels of membrane fatty acid desaturation are important in chill sensitivity, which was suggested by the earlier correlative data.

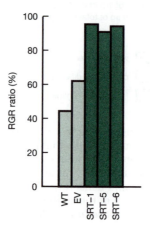

Figure 7.6 Growth characteristics of the wild-type (WT) and transgenic tobacco plants transformed with either the 'empty' vector (EV) or the *Arabidopsis* ω–3 fatty acid desaturase gene (SRT-1, SRT-5 and SRT-6). RGR, relative growth rate at 25°C for 5 days following a 7-day exposure to 1°C. RGR expressed as the ratios of the growth rate of control plants grown at 25°C for 5 days at the same developmental stage. (After Kodama, H., Hamada, T., Horiguchi, G., Nishimura, M. and Iba, K. (1994) Genetic enhancement of cold tolerance by expression of a gene for chloroplast ω–3 fatty acid desaturase in transgenic tobacco. *Plant Physiology*, **105**, 601–605.)

Table 7.1 Fatty acid composition of whole leaves from wild-type and transgenic tobacco plants transformed with either the 'empty' vector or with the *Arabidopsis* ω-3 fatty acid desaturase gene (SRT-1, SRT-5, SRT-6). The measurements were made on the transgenic progeny (seedlings) of the primary transformants, that is on plants that had inherited the transgene. The values are mol% ± SD.

Fatty acid	Wild type	'Empty' vector	SRT-1	SRT-5	SRT-6
16:2	1.5 ± 0.4	1.8 ± 0.2	0.2 ± 0.2	trace	0.4 ± 0.3
18:2	19.5 ± 0.6	20.1 ± 0.8	14.4 ± 0.2	15.7 ± 0.7	14.2 ± 0.5
16:3	7.9 ± 0.9	7.5 ± 0.3	11.3 ± 0.2	8.7 ± 1.1	13.2 ± 1.5
18:3	53.3 ± 0.8	51.3 ± 0.9	59.2 ± 0.7	58.8 ± 0.2	56.6 ± 0.9

Addition of new control sequence

Genes involved in the control of gene expression have also been manipulated by Ti plasmid transformation.

The *R* gene of maize controls the production of anthocyanin pigments. It encodes a transcriptional activator protein that has an acidic and a basic domain with homology to the helix–loop–helix DNA-binding and dimerization domains of some mammalian transcriptional regulators or factors. These transcriptional regulators can act in *trans*, that is the gene does not have to be adjacent to the gene it is controlling.

The *R* gene from maize has been cloned and transformed into tobacco (cv. Xanthi) under the control of the constitutive CaMV 35S promoter. The results show that in comparison with flowers from the non-transgenic parent the transformant shows elevated levels of anthocyanin pigment in the transgenic flowers. This experiment illustrates a number of points:

1. the monocotyledon transcriptional activator works in a dicotyledon;
2. the levels of native tobacco transcriptional activator must be limiting;
3. in the tobacco cultivar used anthocyanin is only produced in the flowers and the tissue distribution of anthocyanin was the same in the transgenic plants, indicating that *R* does not influence the spatial pattern of gene expression (NB in other systems anthocyanin can be synthesized in vegetative parts of plants).

Use of reporter gene constructs

The coding sequence of an easily measured non-plant protein can be used to monitor the ability of *cis*-acting sequences (promoters) to control gene expression. These sequences are called reporter genes and the most commonly used is a bacterial β-glucuronidase gene (GUS; see section 7.4). The assay for β-glucuronidase activity is destructive and alternative reporters, such as the firefly luciferase gene or a green fluorescent protein gene from a jellyfish, can be used if a non-destructive assay is required.

In an analysis of the *Arabidopsis* alcohol dehydrogenase gene (*Adh*) promoter a 1-kb genomic fragment of the putative promoter region (–964 to +53) was joined to the coding sequence of a GUS reporter gene (Figure 7.7a). Transgenic *Arabidopsis* plants containing this chimeric gene were produced by *Agrobacterium* Ti plasmid transformation. Figure 7.7b shows that GUS

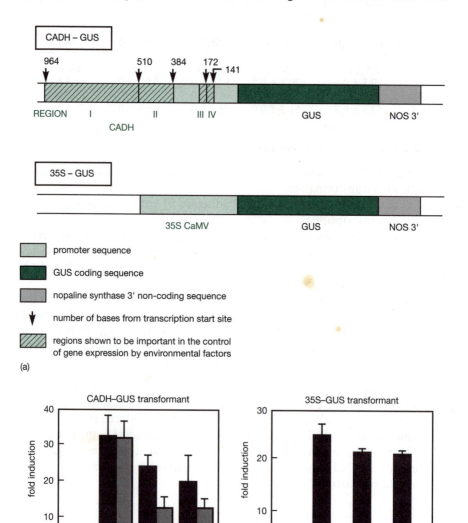

Figure 7.7 Analysis of the *Arabidopsis* alcohol dehydrogenase (ADH) gene promoter using the β-glucuronidase (GUS) reporter gene. (a) Reporter gene constructs: CADH, *Arabidopsis* alcohol dehydrogenase gene promoter; 35S, CaMV cauliflower mosaic virus promoter; GUS, β-glucuronidase. (b) Bar diagrams showing-fold induction of GUS mRNA in transformed plants compared with endogenous ADH mRNA. Solid black bars, ADH mRNA induction levels; hatched bars, GUS mRNA induction levels. AN, low oxygen; C, control; Cd, low temperature; D, dehydration. (After Dolferus, R., Jacobs, M., Peacock, W.J. and Dennis, E.S. (1994) Differential interactions of promoter elements in stress responses of the *Arabidopsis Adh* gene. *Plant Physiology*, **105**, 1075–1087.)

mRNA is induced by the same environmental factors that induce the endogen-ous ADH mRNA, whereas a control construct containing the 35S CaMV pro-moter (see Figure 7.7a) shows no GUS mRNA induction by the environmental treatments, low oxygen, dehydration and cold.

Deletions of the *Adh* promoter were then tested for their ability to confer low oxygen-, dehydration- and low temperature-induced expression of GUS and the results are shown in Table 7.2. These results show that a number of regions of the 964-bp promoter fragment are involved in the environmental control of a GUS reporter. Region I (Figure 7.7a) appears to have a negative regulatory role since removing this leads to a general increase in expression. Region II is impor-tant as a positive regulatory region in the general level of expression and regions III and IV appear to be particularly important in environmental induction since the levels of expression in low oxygen, dehydration and low temperature are affected more by the deletions than the level of uninduced expression.

This type of analysis can be followed by a more detailed analysis using site-directed mutagenesis of individual bases or small groups of bases within the regions of interest, followed by transformation with the modified reporter con-struct and further expression studies.

Table 7.2 Analysis of 5' deletion mutants of *Adh* promoter for control of gene expression by environmental factors. CADH-GUS, full promoter (964 bp); Δ-510, 510 bp of promoter, region 1 deleted; Δ-384, 384 bp of promoter, regions 1 and 2 deleted; Δ-172, 172 bp of promoter, regions 1, 2 and 3 deleted; Δ-141, 141 bp of promoter, regions 1, 2, 3 and 4 deleted

	Relative GUS expression levels (%)			
Construct	**Uninduced**	**Low O$_2$**	**Dehydration**	**Cold**
CADH-GUS (−964)	100	100	100	100
ADH-GUS Δ-510	185	283	223	478
ADH-GUS Δ-384	44	14	47	46
ADH-GUS Δ-172	14	2	3	3
ADH-GUS Δ-141	28	1	5	15

Removal of gene function

The removal of an existing gene function from a plant is more difficult than the addition of a gene function, which is the basis of the previous three examples. However a number of successful examples of removing gene function, based on the strategy of using antisense gene constructs, have been reported in plants.

Figure 7.8 explains the basis of this technique where the coding sequence of the chosen gene is spliced in the reverse orientation between a promoter sequence and a normal transcription termination region. During transcription of a normal gene, RNA polymerase copies one of the DNA strands, reading from the promoter in the 3' to 5' direction (the lower strand of the DNA duplexes in Figure 7.8). The mRNA sequence is then equivalent to the 5'–3' DNA strand (upper strand in Figure 7.8). In the antisense construct, transcription also copies the 3'–5' DNA strand but during the reversal of the sequence and splicing, the strands must be exchanged. During transcription, therefore, RNA polymerase copies the original 'upper' DNA strand in the 3' to 5' direction producing an mRNA that is complementary to the sequence of the normal mRNA. The position of the the the first ATG codon in the normal gene and in the antisense construct is shown in Figure 7.8 to help understand the orientation of the nucleic acid sequences.

The precise mechanism by which the addition of an antisense construct down-regulates the expression of the endogenous gene in transgenic plants is not known. An attractive proposal is shown in Figure 7.8c. Since the sense mRNA from the normal endogenous gene and the antisense mRNA from the transgene are complementary, they may hybridize to form a double-stranded RNA molecule. This molecule may be unstable and rapidly degraded since transgenic antisense plants commonly contain reduced levels of both types of mRNA molecules.

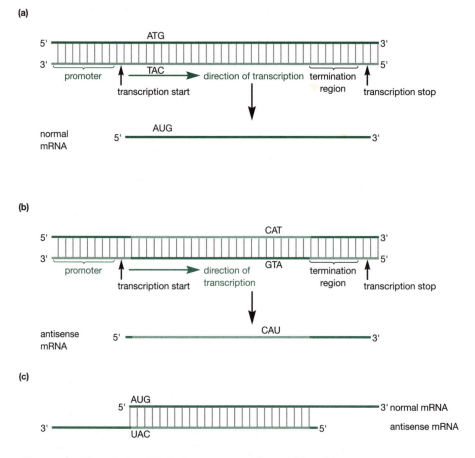

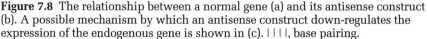

Figure 7.8 The relationship between a normal gene (a) and its antisense construct (b). A possible mechanism by which an antisense construct down-regulates the expression of the endogenous gene is shown in (c). I I I I, base pairing.

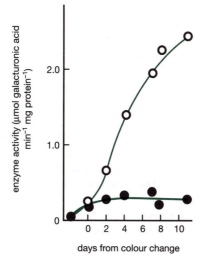

Figure 7.9 Production of polygalacturonase during tomato fruit ripening in normal and antisense transgenic plants. O, normal (control) fruit; ●, antisense transgenic fruit. (After Smith, C.J.S., Watson, C.F., Ray, J., Bird, C.R., Morris, P.C., Schuch, W. and Grierson, D. (1988) Antisense RNA inhibition of polygalacturonase gene expression in transgenic tomatoes. *Nature*, **334**, 725–726.)

The use of antisense constructs is best known in studies of tomato ripening. Figure 7.9 shows the production of the enzyme polygalacturonase in tomato fruits from non-transgenic control plants and in fruit of tomato plants containing the antisense of the polygalacturonase gene. Inhibition of the rise of polygalacturonase activity during ripening can be in the antisense transgenic plants. The development of colour during ripening is normal in the transgenic fruit but because polygalacturonase is involved in softening of the cell walls during ripening, the transgenic fruit remain firm for longer than the normal fruit. This technique is the basis of the 'Flavr Savr' tomato, which has a longer shelf-life and is on sale in supermarkets in the USA (see Part 4 for discussions of the development of this technology).

7.6 Co-suppression

One phenomenon often seen in transgenic plants, for which there is still no accepted explanation, is co-suppression or gene silencing. This is generally seen to occur when multiple copies of homologous sequences arise in a plant following transformation. The transgenes are not antisense constructs and the silencing or co-suppression can occur between the endogenous gene and the homologous transgene or between multiple copies of transgenes. This phenomenon is not restricted to homologous coding sequences but may also occur with promoter sequences or with both.

In some cases the inactivation is considered to be non-reciprocal, in the sense that one 'silent' copy of the gene imposes inactivation on another gene. This type of interaction probably involves methylation whereby a methylated sequence is able (by an unknown mechanism) to impose a methylation pattern on a second copy of the same sequence.

Co-suppression, in contrast, is a reciprocal process where both copies of the sequence are silent (not expressed) as a consequence of the presence of both in the same genome. Some examples of co-suppression are thought to be due to a RNA degradation process, which is triggered when mRNA levels reach a certain value. In other examples, particularly where promoters are involved, this is unlikely to be an explanation and here nucleic acid interactions, either RNA–DNA or DNA–DNA, are thought to bring about *de novo* DNA methylation, which prevents expression.

The phenomenon of co-suppression has been studied in *Petunia* where the addition of extra copies of the *Petunia* chalcone synthase (CHS) gene had the surprising effect of suppressing the expression of both the endogenous homologous gene and the transgene. These CHS transgenes were not inserted near to the endogenous CHS gene. The enzyme CHS is involved in anthocyanin pigment biosynthesis and the co-suppression produces striking transgenic plants with white flowers.

Major Learning Objectives for Chapter 7

1. Knowledge of the method of producing transgenic plants using a Ti plasmid binary vector system.
2. Be aware of the constraints on heterologous (foreign) gene expression in transgenic plants.
3. Understand the analytical use of plant transformation in molecular plant studies, and in particular:
   (a) be acquainted with the use of reporter genes,
   (b) be able to explain antisense technology.

Further reading

The journal *Plant Cell and Environment* produced a special issue (5) in vol. 17 (1994) which is largely devoted to review articles on the use of transgenic plants in experimental studies of a wide range of plant processes. Included in this issue are the following three references:

GRAY, J.E., PICTON, S., GIOBANNONI, J.J. and GRIERSON, D. (1994). The use of transgenic and naturally occurring mutants to understand and manipulate tomato fruit ripening. *Plant Cell and Environment*, **17**, 557–571.

LEA, P.J. and FORDE, B.G. (1994). The use of mutants and transgenic plants to study amino acid metabolism. *Plant Cell and Environment*, **17**, 541–556.

WHITELAM, G.C. and HARBERD, N.P. (1994). Action and function of phytochrome family members revealed through the study of mutant and transgenic plants. *Plant Cell and Environment*, **17**, 615–626.

MATZKE, M.A. and MATZKE, A.J.M. (1995). How and why do plants inactivate homologous (*Trans*) genes? *Plant Physiology*, **107**, 679–685.
This review paper discusses three possible mechanisms of gene silencing in more depth than possible in this chapter.

WILLMITZER, L. (1988). The use of transgenic plants to study plant gene expression. *Trends in Genetics*, **4**, 13–18.
A description of the analysis of promoters (*cis*-regulatory sequences) using reporter genes.

Part three

Plant molecular biology

Chapter 8

Symbiotic nitrogen fixation

8.1 Biology of nitrogen fixation

Nitrogen (N_2) constitutes 78% of the air we breathe in and 78% of the air we breathe out. In other words, we cannot convert this very stable gas into a biologically useful form. Prokaryotes are the only organisms that can fix atmospheric N_2 into a compound able to be metabolized. N_2-fixing prokaryotes can be grouped as follows:

Bacteria
    Non-photosynthetic, non-filamentous
        Free living, e.g. *Klebsiella* spp.
        Symbiotic, e.g. *Rhizobium* spp.
    Photosynthetic, e.g. *Rhodospirillum*
    Filamentous (actinomycetes), e.g. *Corynebacterium* spp.
    Cyanobacteria, e.g. *Anabaena* spp.

All of these organisms reduce N_2 to ammonia:

$$N_2(g) + 3H_2 (g) \rightarrow 2NH_3(g) \quad \Delta H = -33.3 \text{ kJ mol}^{-1}$$

This reaction is exothermic and produces 33.3 kJ mol^{-1}, but it does not occur spontaneously at normal temperature and pressure because of the enormous stability of the N_2 molecule. N_2-fixing organisms produce an enzyme complex (nitrogenase–nitrogenase reductase) that is able to reduce N_2 to NH_3, but in all organisms this enzyme is irreversibly inactivated by oxygen (O_2). This presents a problem for N_2-fixing cells since the reaction requires energy and is therefore dependent on the chemical energy source ATP. ATP may be generated in cells from:

1. photophosphorylation in photosynthetic cells, but this process produces O_2;
2. oxidative phosphorylation in all cells with a suitable terminal electron acceptor, but O_2 is the most widespread of these;
3. substrate-level phosphorylation in cells that are anaerobic.

Many of the free-living organisms that fix N_2 only do so under anaerobic conditions. However, the symbiotic bacteria solve the problem in collaboration with a plant by controlling the supply of O_2. The most widely studied and the most important symbiotic relationship is that between bacterial *Rhizobium* spp. and plant species of the family Leguminosae, and this chapter will be limited to these examples.

8.2 Development of legume nodules

The development of a legume root nodule is illustrated in Figure 8.1. *Rhizobium* spp. are free-living soil bacteria able to invade legume roots and cause the development of outgrowths, called nodules. There is specificity of invasion and nodule formation, such that a given species (or biovar) of *Rhizobium* is only capable of a symbiotic relationship with one (or a few) plant species. Table 8.1 shows the host range for a number of *Rhizobium* spp.

The first reaction of the plant to the presence of compatible *Rhizobium* in the vicinity of the root (rhizosphere) is curling of the root hairs. The bacteria enter the root via the curled root hairs and induce the formation of an infection thread. This structure (Figure 8.2), which contains the multiplying bacteria, is formed from invaginations into and through the plant cells and is therefore bounded by cell walls and plant cell membranes (plasma membranes). Eventually the bacteria reach the inner root cortex via the proliferating infection thread and here the bacteria are taken up into cells from an unwalled infection droplet by endocytosis (Figure 8.3). The resulting intracytoplasmic bacteria are surrounded by a periplasmic membrane of plant origin. There follows

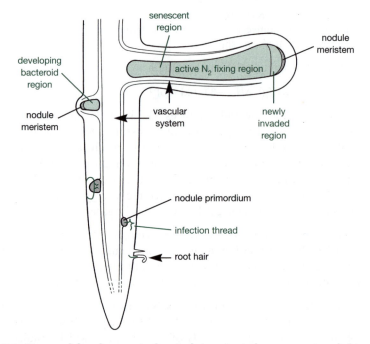

Figure 8.1 Stages of development of an indeterminate legume root nodule.

Table 8.1 *Rhizobium* nodulation host range

Bacterium	Plant host
Rhizobium meliloti	Alfalfa
Rhizobium leguminosarum	
Biovar *viciae*	Pea
Biovar *trifolii*	Clover
Biovar *phaseli*	*Phaseolus* beans
Bradyrhizobium japonicum	Soya bean
Azorhizobium caulinodaus	*Sesbania* (a tropical shrub with nodules on both roots and leaves)

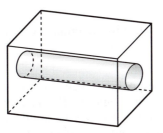

Figure 8.2 Path of an infection thread passing through a root cell.

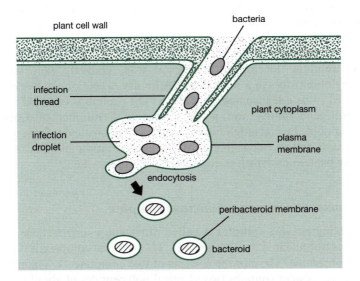

Figure 8.3 Endocytosis of bacteria from an unwalled infection droplet to form symbiotic bacteroids with a peribacteroid membrane of plant origin. (From Brewin, N.J., Ambrose, M.J. and Downie, J.A. (1993) Root nodules, *Rhizobium* and nitrogen fixation. In Casey, R. and Davies, D.R. (eds) *Peas: genetics, molecular biology and biotechnology*, pp. 237–290. CAB International, Wallingford.)

a differentiation of this membrane together with the enclosed bacterium to form the differentiated nitrogen-fixing bacteria, called bacteroids. The cortical cells adjacent to the bacteroid-containing cells become meristematic; proliferation of these cells followed by continuing endocytotic infection by the bacteria results in the formation of the nodule (see Figure 8.1). In some species the nodules retain a meristem throughout their life but in other species the nodules are determinate (have no meristem when mature).

The differentiation of bacteroids involves a number of changes to the bacterial membrane, the production of a number of auto-oxidizable cytochromes as well as the production of the N_2-fixing apparatus. The plant cells also change, so that bacteroid-containing cells have no vacuoles and nodule cells produce a number of specific 'nodule' proteins, the most abundant of which is leghaemoglobin.

8.3 Leghaemoglobin and the nitrogenase complex

Leghaemoglobin

There is an invariable correspondence between the N_2-fixing activity of nodules and the presence of haemoglobin (leghaemoglobin). This is a rare molecule in the plant kingdom but it is structurally very similar to the animal haemoglobin. The two components of this protein are produced by different symbionts, with the bacteroids producing the haem molecule co-factor and the plant producing the apoprotein (polypeptide). Leghaemoglobin accumulates to very high levels in the legume root nodules, where it can represent 30% of the total nodule protein and 20% of the mRNA. The protein gives active legume root nodules a characteristic pink colour.

Legume leghaemoglobin apoprotein genes were among the first nuclear plant genes to be cloned. Since leghaemoglobin mRNA forms 20% of the mRNA found in developing nodules it was relatively easy to identify leghaemoglobin clones in a nodule cDNA library, where one in five clones can be expected to be a leghaemoglobin clone. The protein is encoded by a small plant multigene family and nodules contain a number of isoforms of leghaemoglobin. Leghaemoglobin is located in the host cytoplasm and not in the bacteroids and its role is to control the concentration and flux of O_2 available to the bacteroids, which contain the N_2-fixing enzyme.

Nitrogenase–nitrogenase reductase complex

The N_2-fixing enzyme complex is called nitrogenase–nitrogenase reductase (sometimes abbreviated to nitrogenase). This enzyme complex is encoded by genes in *Rhizobium*. The enzyme consists of two components. The nitrogenase reductase component contains two identical polypeptides of about 60×10^3 M_r each, together with four iron (Fe) and four labile sulphur (S) atoms. The other component, the nitrogenase, is a tetramer, having two molecules each of two different polypeptides (2 + 2). It also contains labile S atoms (24) and Fe ions (12–32), together with two molybdenum (Mo) ions.

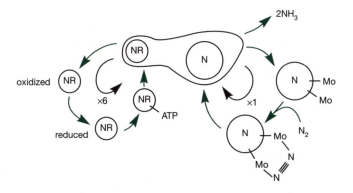

Figure 8.4 Activity of nitrogenase (N) and nitrogenase reductase (NR) in N_2 fixation.

The two components of this enzyme have different functions. Nitrogenase reductase collects reducing power and energy whilst nitrogenase collects and reduces the substrate (N_2), producing NH_3. This reaction is shown diagrammatically in Figure 8.4. Six cycles of nitrogenase reductase activity are required for every N_2 molecule that is reduced. It has been calculated that the process of N_2 fixation is only 80% efficient and that 15 ATP molecules are required for every NH_3 molecule (i.e. NH_4^+ ion) produced.

8.4 N_2 fixation in the free-living bacterium *Klebsiella pneumoniae*

There are clearly special problems involved in studying a process that requires the collaboration of two organisms and therefore N_2 fixation was first studied in the free-living soil bacterium *Klebsiella*. Figure 8.5 shows the *Klebsiella nif* gene cluster, which contains 17 genes in seven to eight operons. The structural genes for the two polypeptides of nitrogenase (*nif*K and *nif*D) and the nitrogenase reductase polypeptide (*nif*H) are present on the same operon, together with a gene or ORF with unknown function. Gene *nif*Q is involved in Mo uptake, genes *nif*B, *nif*V, *nif*N and *nif*E are involved in processing the Mo co-factor and genes *nif*F and *nif*J are probably responsible for electron transport. Two regulatory genes in the same operon have been identified: *nif*A is an activator whereas *nif*L is a repressor.

Box 8.1 Regulation of *nif* gene expression in *Klebsiella*

The regulatory network for the *nif* gene cluster is shown in Figure a. Three genes not present in the *nif* gene cluster (Figure 8.5), *ntr*A, *ntr*B and *ntr*C, are involved in the regulation of *nif* genes. The *ntr*A gene product is required for the recognition of *nif* promoters by RNA polymerase. For transcription of the *nif*LA regulatory operon, the phosphorylated form of the *ntr*C gene product is also required and the phosphorylation of this protein is controlled by N_2 levels via the *ntr*B gene product. Thus in N_2-limited media, NtrC (the protein encoded by *ntr*C) is phosphorylated and *nif*L and *nif*A are expressed. The ▸

(Box 8.1 continued)

The *nif*A gene product is a transcriptional activator of other *nif* genes. The *nif*L gene is a negative regulator; however the *nif*L gene product can only inhibit *nif*A activation in the presence of fixed N_2 or excess O_2.

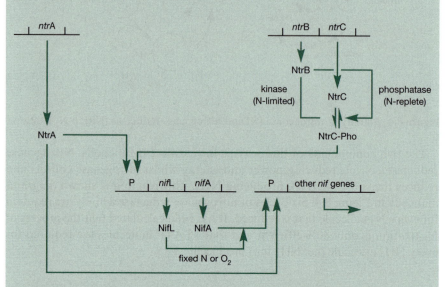

Figure a Regulation of *nif* gene expression in *Klebsiella*: NtrA, *ntr*A gene product; NtrB, *ntr*B gene product; NtrC, *ntr*C gene product; NtrC-Pho, phosphorylated NtrC protein; NifA, *nif*A gene product; NifL, *nif*L gene product; P, promoter. (From Smith, R.J. and Gallon, J.R. (1993) Nitrogen fixation. In Lea, P.J. and Leegood, R.C. (eds) *Plant biochemistry and molecular biology*, p. 129. J.Wiley, Chichester.)

8.5 *Rhizobium* genes for symbiotic N_2 fixation and nodulation

The genes for N_2 fixation and nodulation in *Rhizobium* spp. are situated on large plasmids (of about 200 kb) and not on the bacterial chromosome. Homologues of some of the *Klebsiella nif* genes have been identified in *Rhizobium*, together with two other classes of genes. These are the *nod* genes, which are involved in nodulation, and the *fix* genes, which have a role in N_2 fixation but are only found in *Rhizobium*. Not all of the plasmids found in *Rhizobium* spp. are the same, with the major differences being in those sequences involved in host range and in the arrangement of the gene clusters on the plasmid.

Figure 8.6 shows the arrangement of genes on the plasmid pRL1JI in *Rhizobium leguminosarum* together with a list of the functions or features known for the gene products. The *nif* genes are all homologues of the equivalent

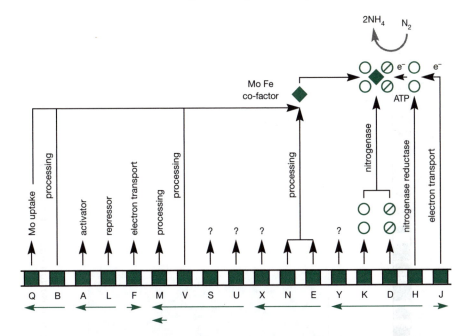

Figure 8.5 Map of the *Klebsiella nif* gene cluster showing the function of each gene (where known) in the assembly and activity of the nitrogenase–nitrogenase reductase complex. (From Vance, C.P. and Griffith, S.M. (1990) The molecular biology of nitrogen fixation. In Dennis, D.T. and Turpin, D.H. (eds) *Plant physiology and molecular biology*, pp. 373–388. Longman, Harlow.)

genes in *Klebsiella*. Thus *nif*K and *nif*D are structural genes for the two nitrogenase polypeptides and *nif*H encodes the nitrogenase reductase protein. Genes *nif*B and *nif*E are involved in the production of the Mo co-factor and *nif*A is a regulatory gene. The genes *fix*A, *fix*B and *fix*C encode proteins that have a role in donating electrons to nitrogenase and *fix*L and *fix*J are thought to have control functions; these genes are not present in *Klebsiella*.

Rhizobium strains with mutations in the *nif* and *fix* genes are able to induce nodule formation in the legume host but these nodules do not fix N₂. The *nod* genes, on the other hand, are characterized by the fact that mutations in these genes lead to the loss of ability to form nodules. The *nod* genes C, L, M, B, F and E are responsible for the formation of lipo-oligosaccharides. The functions of these genes are shown in Figure 8.6b, while Figure 8.7 illustrates the structure of the lipo-oligosaccharide secreted by *R. leguminosarum* biovar *viciae* and the probable role of several *nod* genes in the biosynthesis of this molecule. The lipo-oligosaccharides are secreted by free-living *Rhizobium*, and induce plant root hair curling. They are also thought to be secreted after invasion, when they induce cell division in root cortical cells. The *nod*D gene is known to have an important regulatory function,

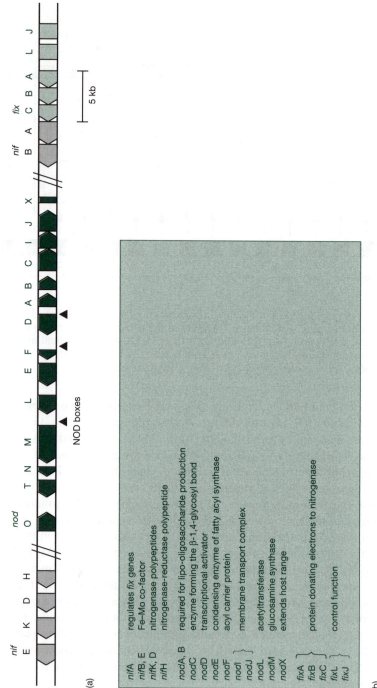

(a)

(b)

nifA   regulates *fix* genes
nifB, E   Fe-Mo co-factor
nifK, D   nitrogenase polypeptides
nifH   nitrogenase-reductase polypeptide

nodA, B   required for lipo-oligosaccharide production
nodC   enzyme forming the β-1,4-glycosyl bond
nodD   transcriptional activator
nodE   condensing enzyme of fatty acyl synthase
nodF   acyl carrier protein
nodI   ⎫
nodJ   ⎬ membrane transport complex
nodL   acetyltransferase
nodM   glucosamine synthase
nodX   extends host range

fixA   ⎫
fixB   ⎬ protein donating electrons to nitrogenase
fixC   ⎭
fixL   ⎫
fixJ   ⎬ control function

Figure 8.6 (a) Map of the *nod*, *nif* and *fix* genes on the *Rhizobium leguminosarum* plasmid pRL1JI;(b) key to the functions where known.

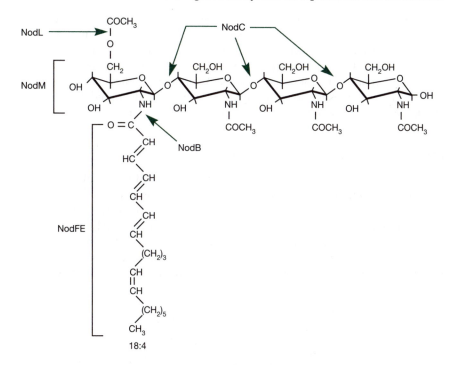

Figure 8.7 Structure of a lipo-oligosaccharide secreted by *Rhizobium leguminosarum* bv. *viciae* showing the probable role of several *nod* genes in the biosynthesis of this nodulation factor. NodC, enzyme forming β-1,4-glycosyl bond; NodL, acetyltransferase; NodM, glucosamine synthase; NodB, *N*-acetylglucosamine deacetylase; NodF, acyl carrier protein; NodE, condensing enzyme of fatty acyl synthase.

8.6 Interaction between bacteria (*Rhizobium*) and plant (legume)

Pre-infection signals

Details of the interaction between bacteria and plant cells in the formation of N$_2$-fixing nodules are beginning to be discovered. There is a series of signal exchanges between the plant and the free-living bacteria, which involve the bacterial lipo-oligosaccharide described in the previous section.

Legume roots secrete flavonoid compounds into the rhizosphere, with a single plant species secreting a number of these compounds. An example of a flavonoid secreted by *Vicia* roots is shown in Figure 8.8. These flavonoids can activate the product of the *nod*D gene on the *Rhizobium* plasmid. The *nod*D gene product is a transcriptional activator and is required for the expression of the other *nod* genes including the genes for lipo-oligosaccharide synthesis. A conserved motif, known as the NOD box (Figure 8.6), has been identified in putative promoter regions within the *nod* gene cluster and it is thought that this NOD box is involved in the regulation of *nod* genes by the *nod*D protein.

It is probable that flavonoids and lipo-oligosaccharides are not the only compounds involved in the pre-invasion signal exchanges, since bacterial succinoglycans are also thought to be important in the invasion of the root by bacteria. In addition, the host-range specificity of *Rhizobium* invasion is unlikely to be fully explained by these compounds. A summary of these pre-invasion and early invasion signals is shown in Figure 8.9 where the flavonoid–lipo-oligosaccharide signal path is numbered to show the sequential steps.

Interaction between bacterial and root nodule cell

At one level the interaction between the bacteroid and plant cell involves an exchange of metabolites. Thus the plant provides the bacteroids with dicarboxylic acids, such as malate and succinate, as substrates for oxidative phosphorylation and the bacteroids provide the plant with NH_4^+ from N_2 fixation. This NH_4^+ is then assimilated by the plant using glutamine synthase to produce glutamine from glutamate and NH_4^+. The plant also provides the Mo, Fe and S necessary for the formation of the active nitrogenase complex.

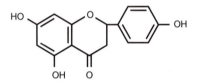

Figure 8.8 The chemical structure of naringenin, an example of a flavonoid compound secreted by *Vicia* roots, which acts as an inducer of *nod* gene transcription in *Rhizobium leguminosarum* bv. *viciae*.

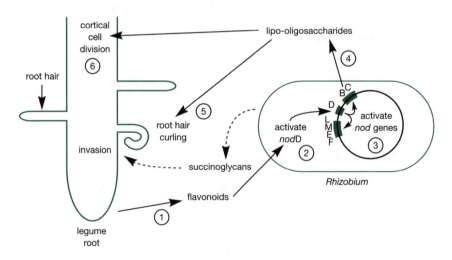

Figure 8.9 Summary of the pre-invasion and early invasion signals between *Rhizobium* and legume root. The numbers refer to sequential steps in the flavonoid–lipo-oligosaccharide signal pathway.

However, at another level, the plant and bacteroid cells interact to provide a favourable environment for N_2 fixation and this involves signals that control the expression of both plant and bacteroid genes. *Rhizobium* invasion results in the formation of 18–20 new plant polypeptides during the formation of the nodules. Functions have been assigned to only a few of these. For example, glutamine synthase levels increase dramatically during nodule development and the plant-encoded apoprotein of leghaemoglobin is only produced in nodule cells. The nature of the signals responsible for these changes in plant gene expression is not known.

Because the nitrogenase complex is inactivated by O_2, levels of O_2 are critical for N_2 fixation and leghaemoglobin controls the flux of O_2 to the bacteroid (see section 8.3). The synthesis of the nitrogenase complex is also controlled by O_2 via the genes *fixL* and *fixJ*: *fixL* encodes a haem protein that senses O_2 and phosphorylates the protein encoded by *fixJ*. The phosphorylation of *fixJ* protein controls its ability to activate *nif*A, which in turn is a transcriptional activator of the other *nif* genes. An overview of the known interactions between bacteroid and plant cell is shown in Figure 8.10.

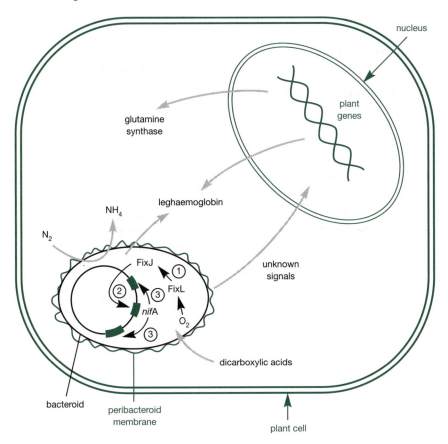

Figure 8.10 An overview of the known interactions between *Rhizobium* bacteroids and legume root cells during N_2 fixation in nodules. The numbers refer to sequential steps in the control of *nif* gene expression by O_2 via the *fixL* and *fixJ* gene products.

Major Learning Objectives for Chapter 8

1. Outline the development of legume N_2-fixing nodules and understand the interaction between *Rhizobium* bacteria and plant roots in the early stages of nodule formation.
2. Describe the structure and function of the nitrogenase–nitrogenase reductase complex and give reasons why N_2-fixing legume nodules produce haemoglobin.
3. Knowledge of the function of the three classes of *Rhizobium* pRL1JL plasmid genes (*nif*, *fix* and *nod*).
4. Explain the metabolic interactions between the bacteroids and plant nodule cells.
5. Understand the similarities and differences between the control of N_2 fixation by O_2 and N_2 in the free-living bacterium *Klebsiella* and the symbiotic *Rhizobium* bacteroids.

Further reading

The following three references, in multi-author texts, more or less cover the material of this chapter. Brewin *et al.* (1993) is based on peas, Smith and Gallon (1993) includes nitrogen fixation in cyanobacteria, and Vance and Griffith (1990) includes information about ammonia assimilation.

BREWIN, N.J., AMBROSE, M.J. and DOWNIE, J.A. (1993). Root nodules, *Rhizobium* and nitrogen fixation. In Casey, R. and Davies, D.R. (eds) *Peas: genetics, molecular biology and biotechnology*, pp. 237–290. CAB International Wallingford.

SMITH, R.J. and GALLON, J.R. (1993). Nitrogen fixation. In Lea, P.J. and Leegood, R.C (eds) *Plant biochemistry and molecular biology*, pp. 129–153. John Wiley & Sons, Chichester.

VANCE, C.P. and GRIFFITH, S.M. (1990). The molecular biology of N metabolism. In Dennis, D.T. and Turpin, D.H. (eds) *Plant physiology, biochemistry and molecular biology*, pp. 373–388. Longman, Harlow.

FISHER, R.F. and LONG, S.R. (1992). *Rhizobium*–plant signal exchange. *Nature*, **357**, 655–660.
This is a review article dealing with the initial stages of *Rhizobium*–legume interaction.

Chapter 9

Tissue-specific expression of plant genes: seed storage protein genes

9.1 Seed storage proteins

The production of functional proteins in plant cells involves a number of different processes (see Figure 1.6) and in this chapter studies of regulatory promoters (*cis*-regulatory sequences) and transcription factors will be illustrated with reference to seed storage proteins. Of the edible protein produced in the world 70% comes from seeds and these proteins are therefore of considerable economic importance. They are described in this chapter partly because of their economic importance and partly because they represent genes that are both temporally and spatially controlled within the plant. In addition, the seed storage proteins of maize and peas are compartmentalized within the plant cells and as such they illustrate further examples of intracellular trafficking (see Chapters 3 and 4).

Seed proteins can be divided into two classes: (i) metabolic proteins, which are also present in the rest of the plant, and (ii) storage proteins, which are unique to the seed. The storage proteins are synthesized in the developing seed and stored for use during germination. They were first named and identified, at the end of the last century, by their successive solubility in a series of solvents:

Solvent	Protein
H_2O	Albumins
Dilute salt	Globulins
Ethanol	Prolamins
Dilute alkali	Glutelins

The major components of legume seed storage proteins are globulins and albumins whereas in cereals the major components are prolamins and glutelins. The synthesis of seed storage proteins is very tightly controlled and these proteins are only synthesized in the developing seed during the later stages of seed development.

Mammals are unable to synthesize a number of amino acids. These are known as the essential amino acids and are commonly supplied in the diet by eating plant proteins.

Box 9.1 Essential amino acids

Leucine
Isoleucine
Lysine
Methionine
Phenylalanine
Threonine
Tryptophan
Tyrosine
Valine

The globulins in legume seeds make up between 20 and 25% of the seed dry weight; however the balance of amino acids in these proteins is not perfect for a mammalian diet because they are deficient in the sulphur-containing amino acids methionine and cysteine. In cereals, the prolamins and glutelins typically make up 8 and 15% of the seed dry weight. These proteins also have an unbalanced amino acid content for mammalian diets and are typically limiting for lysine, threonine and tryptophan.

In view of their importance as food, seed storage proteins have been the subject of a great deal of molecular investigation. They are also particularly difficult to study at the protein level, partly because of the large number of different proteins produced by any individual plant and partly because of problems arising from the solubility characteristics of some storage proteins. Molecular studies have therefore been able to provide fundamental information about the proteins themselves, as well as the genes that encode them or control their synthesis. One aspect of storage proteins that has received considerable attention is the possibility of altering the amino acid composition of these proteins in order to improve their nutritional quality.

In this chapter two examples of storage proteins will be covered: the zein (prolamin) storage proteins of maize and the legumin and vicilin (globulins) storage proteins of pea.

9.2   Zein proteins of maize (*Zea mays*)

Zeins are synthesized in the cytoplasm by membrane-bound polyribosomes in the developing maize (monocotyledon) endosperm. The proteins are co-translationally transported into the lumen of the rough endoplasmic reticulum (RER) where they aggregate into insoluble masses called protein bodies.

Zein synthesis occurs during the period between free nuclear division within the endosperm and desiccation of the seed. At 16 days after pollination it has been calculated that in each endosperm zein proteins are being synthesized at 300 µg day^{-1}. Figure 9.1 shows the accumulation of zein proteins during the period from 10 to 28 days after pollination. The zein proteins were extracted in an alcohol solution and separated by sodium dodecylsulphate (SDS) polyacrylamide gel electrophoresis

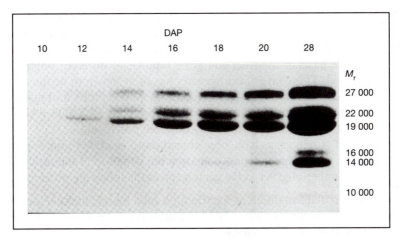

Figure 9.1 Sodium dodecylsulphate polacrylamide gel electrophoretic analysis of zein proteins from developing endosperm of maize. The proteins were extracted with 70% ethanol containing 2% 2-mercaptoethanol. DAP, days after pollination; M_r, relative molecular mass. (From Larkins, B.A., Lending, C.R. and de Barros, E. (1991) Assembly of maize storage proteins into protein bodies in developing endosperm. In Hermann, R.G. and Larkins, B.A. (eds) *Plant molecular biology*, vol. 2, pp. 619–625. Plenum Press, New York.)

(PAGE), which separates proteins solely on the basis of their relative molecular mass (size). Using this technique several forms of zein can be distinguished:

$M_r = 27\,000$ γ-zein   S-rich
$M_r = 22\,000$ α-zein
$M_r = 19\,000$ α-zein
$M_r = 16\,000$ γ-zein   S-rich
$M_r = 14\,000$ β-zein   S-rich
$M_r = 10\,000$ δ-zein   S-rich

The, α, β and γ zeins differ sufficiently in structure for group-specific antibodies to not cross-react with the other groups. Using these antibodies to localize the different groups of zein in protein bodies, it has been shown that they are not deposited uniformly (Figure 9.2).

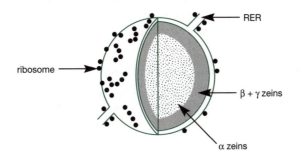

Figure 9.2 Structure of a maize endosperm protein body containing zein storage proteins. RER, rough endoplasmic reticulum.

The 22 000 M_r and 19 000 M_r zeins (α-zeins) make up about 70–80% of the maize storage proteins. When an individual molecular weight class of zein is separated with isoelectric focusing (IEF) gel electrophoresis (which separates proteins on the basis of their net charge) a number of different polypeptides can be identified. This charge heterogeneity has been shown to be the result of amino acid composition differences between the polypeptides and not caused by post-translational modification of amino acids. Thus each SDS PAGE zein band consists of a number of structurally different polypeptides.

A number of α-zein genes have been cloned, both as cDNA and as genomic DNA, and the use of these clones to analyse the structure of the proteins and estimate the number of genes in the genome gives the following picture.

1. The two α-zeins of different relative molecular mass (22 000 M_r and 19 000 M_r) are structurally related but there is greater sequence homology between α-zeins of the same size.
2. The 22 000 M_r α-zeins have 90% sequence homology and are encoded by a multigene family with about 25 members, some of which are inactive (sometimes called pseudogenes).
3. The 19 000 M_r α-zeins can be divided into two subfamilies on the basis of sequence similarity. This group is encoded by a family of about 55 members, again with some of the genes appearing to be inactive.
4. The genes contain no introns.
5. The α-zein amino acid sequence predicted from the cDNA sequence is larger than that calculated from SDS PAGE. This is because the primary polypeptide includes an N-terminal 21-amino acid signal sequence that is responsible for transport of the protein across the endoplasmic reticulum to the lumen, where the accumulated proteins form protein bodies. This signal peptide is removed during transport resulting in a 5 000 M_r reduction in the size of the mature protein.
6. Both α-zeins contain a central domain with a 20-amino acid sequence that is tandemly repeated.

The 20-amino acid repeats are predicted to form α-helical secondary structures that could stack in an antiparallel configuration. This teriary structure of the α-zein is thought to be important in the deposition of high concentrations of protein in the protein bodies. It may also be important in the preservation of the protein during dehydration in such a way that it is able to be degraded by proteases during rehydration and seed germination.

Box 9.2 Number of genes encoding , β, γ and δ-zeins

It is estimated that the γ-zeins are encoded by a small multigene family of about six members, whereas the β and δ-zeins are each thought to be encoded by a single gene.

Using IEF variation in zein proteins, the structural genes for 20 α-zeins have been mapped by classical genetics. The genes are distributed as follows:

chromosome 4 short arm   8 genes
chromosome 4 long arm   2 genes
chromosome 7 short arm   9 genes
chromosome 10 long arm   1 gene

Although most of the zein genes are situated on chromosomes 4 and 7, they are not clustered into one region of these chromosomes.

9.3 Identification of a *trans*-acting factor controlling zein synthesis in maize

In addition to the structural genes for zein proteins, a number of genes have been identified that affect the synthesis of zein proteins and some of these are listed in Table 9.1. This table shows that these genes are also present on chromosomes 4, 7 and 10. However, although there is linkage between some of the zein structural genes and these controlling genes, the control function is not limited to the linked structural gene. This implicates *trans*-acting regulatory factors in the mechanism of control by these genes.

Table 9.1 also illustrates the complexity of the control system for these genes because it shows that mutations at a number of genes can affect zein synthesis. Some of the mutations are recessive (cf. *opaque*-2) whilst others are semi-dominant or dominant. In addition some mutations affect all of the zein structural genes (cf. *opaque*-6 and *floury*-2) whilst others only affect a particular class of zein (cf. *opaque*-2 and *opaque*-7). The interaction of these genes also varies; thus *opaque*-2, *opaque*-6 and *opaque*-7 are additive whereas *opaque*-2 mutations are epistatic (prevent the expression) over *floury*-2. *Opaque*-6 is lethal to seedlings when homozygous and *opaque*-7 affects other endosperm characters; therefore these 'control' genes may not be specific for zein proteins. In *Defective-endosperm*-B30 mutants the synthesis of the 22 000 M_r zeins is delayed by about 10 days and in *opaque*-2 the synthesis of these zein polypeptides stops prematurely. Taken together, these results suggest a complex regulatory network controlling zein synthesis in maize.

We know very little about the molecular basis of this regulatory network. The best characterized control gene is *opaque*-2, which has been cloned by

Table 9.1 Genes affecting zein synthesis

Gene	Chromosome location	Dominance	Per cent inhibition of zein synthesis	Zein protein affected
opaque-2	7 (short)	Recessive	47	22 000 M_r
opaque-6	?	Recessive	88	All
opaque-7	10 (long)	Recessive	77	19 000 M_r
floury-2	4 (short)	Semi-dominant	34	All
Defective-endosperm-B30 (De-B30)	7	Dominant	12	22 000 M_r

transposon tagging with the endogenous maize transposable element Ds. The wild-type allele of *opaque-2* produces a 2-kb transcript encoding a 51 000 M_r polypeptide and *opaque-2* mutants produce no *opaque-2* mRNA (Figure 9.3). Analysis of the predicted *opaque-2* amino acid sequence shows that the protein contains a leucine zipper and zinc finger DNA-binding domains. The sequence has homology with a yeast transcription factor, GCN4, and recently it has been shown that the normal *opaque-2* gene can complement a yeast (*Saccharomyces cerevisiae*) *gcn*4 mutant. This provides experimental confirmation that *opaque-2* encodes a transcription factor that is the *trans*-acting regulatory factor predicted by the genetic analysis of *opaque-2* mutants. The experiment also indicates some conservation of transcription factors between the fungus, yeast and higher plants.

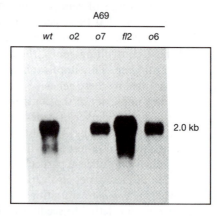

Figure 9.3 Northern blot analysis of *opaque-2* gene expression in the endosperm of wild type and four mutant maize genotypes: *wt*, wild type; *o2*, *opaque-2*; *o7*, *opaque-7*; *fl2*, *floury-2*; *o6*, *opaque-6*. (From Motto, M., Di Fonzo, N., Hartings, H., Maddaloni, M., Salamini, F., Soave, C. and Thompson, R.D. (1989) Regulatory genes affecting maize storage protein synthesis. *Oxford Surveys of Plant Molecular and Cell Biology*, **6**, 87–114.)

9.4   Legumins and vicilin/convicilins in pea (*Pisum sativum*)

The seed storage proteins of the dicotyledon garden pea (*P. sativum*) are composed of globulins and albumins, with the globulins (60%) being the major component. The globulins can be separated into two groups on the basis of the size of the aggregated polypeptides. The legumins have a sedimentation coefficient of 11–12S, whereas the vicilins and convicilins have a sedimentation coefficient of 7S.

The globulin proteins are synthesized in the storage parenchymal cells of the cotyledons (part of the 2*n* embryo) rather than the endosperm as in maize. Similar to maize, however, synthesis occurs during the later stages of seed development and stops during the dehydration stage. At the peak of synthesis globulin mRNA can be 10% of the total seed mRNA population. Also, like maize, the proteins are synthesized in the cytoplasm by ribosomes bound to the endoplasmic reticulum and co-translationally transported into the lumen of the RER. This transport is also accompanied by the removal of an N-terminal signal peptide, which is responsible for this stage of intra-cellular localization.

Legumins and vicilin/convicilins are found in protein bodies in pea cotyle-dons but the protein bodies do not arise simply as protein accumulations within the RER. The proteins accumulate on the vacuolar side of the tono-plast (the membrane which surrounds the vacuole), and as they accumulate the tonoplast appears to evaginate around the protein deposits to pinch off protein bodies. The vacuole gradually gets smaller as more and more pro-tein is deposited and more protein bodies form. The process is also associated with an increase in membrane area. These protein bodies can be shown to contain a number of hydrolytic enzymes that are commonly found in vacuoles.

The site of protein accumulation (the vacuole) and the site of protein synthe-sis (the RER) are spatially separated in the cell and the transport of the proteins within the cell involves the Golgi apparatus. There are direct tubular connections between the RER and the cisternae of the Golgi apparatus. In animal cells sorting of proteins within the Golgi apparatus is associated with glycosylation of the proteins. However, although vicilin is glycosylated legu-mins are not and therefore there are aspects of the intracellular sorting of pea seed storage proteins still not well understood.

Box 9.3 Vacuolar targeting signals

Legumin subunits are also synthesized as precursor polypeptides and transported into protein storage vacuoles in the field bean. In order to study vacuolar targeting signals gene fusions were constructed between segments of the field bean legumin polypeptide and either yeast invertase or chloramphenicol acetyltransferase as reporters. Transport into vacuoles was studied in transformed yeast and transgenic tobacco. The conclusion of this study was that legumin does not contain a single vacuolar targeting signal; rather the polypeptide has multiple areas of targeting information, probably formed by secondary structures of the polypeptide.

Legumin structure

Information from cDNA clones, genomic clones and protein structure indi-cates that there are four classes of pea legumin. These all exist as dimers of structurally different polypeptide subunits (α and β) that are bonded together via a disulphide bridge. The heterodimer is synthesized as a single 60 000 M_r precursor, with the structure NH_2-α-β-COOH. The covalent peptide bond between the α and β polypeptides is thought to be cut after the proteins have been deposited in the protein bodies. The legumin dimers aggregate to form a large oligomeric structure of 360 000–400 000 M_r with a sedimentation coeffi-cient of 11–12S.

Vicilin/convicilin structure

The vicilins are a very heterogeneous group of polypeptides with M_r ranging from about 12 000 to about 70 000. Sequence data from cDNA clones can classify these into five classes. The heterogeneity arises from: (i) existence of gene families, (ii) proteolytic processing after translation and (iii) post-translational glycosylation of polypeptides. The vicilins are synthesised as precursor polypeptides of 47 000–50 000 M_r or about 68 000 M_r. Figure 9.4 shows a diagrammatic representation of the post-translational processing of the 47 000–50 000 M_r vicilin precursor polypeptide to give rise to a range of different mature polypeptides. The M_r of the C-terminal fragment depends on its glycosylation.

Convicilin is equivalent to the 50 000 M_r vicilin polypeptide with a 120–166 sequence of hydrophilic amino acids inserted near the N-terminus.

Pea globulin genes

It has been estimated that there are more than 10 legumin genes in the pea genome. Using qualitative differences in the proteins between different pea cultivars, the structural genes for legumin have been mapped to three loci on chromosome 7. In addition a molecular marker (RFLP) has located one gene on chromosome 1. The only definitive association between any of the cloned genes and an individual legumin protein is that between *leg*A genes and the L4α and β polypeptides. The legumin *leg*A gene contains three small introns.

There are approximately 24 vicilin genes grouped into seven loci on chromosome 7 and there are two convicilin genes on chromosome 2.

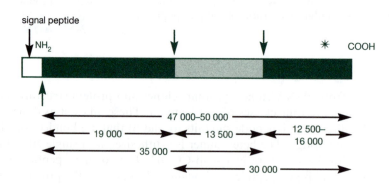

Figure 9.4 Pattern of post-translational processing of the 47 000–50 000 M_r pea vicilin precursor polypeptide to give rise to a range of different mature polypeptides. Asterisk indicates glycosylation site; ↓, proteolytic processing site.

9.5   Identification of *cis*-acting regulatory sequences controlling legumin biosynthesis in pea

The structure of the genomic clone of the pea *leg*A gene is shown in Figure 9.5a. The ATG translation start site is shown together with the exons (solid bars) and the polyadenylation signal. The region upstream of the ATG site has been expanded (Figure 9.5b) to show a number of sequence features:

1.   TATA and CAAT boxes proximal to the translation start ATG;
2.   a 28-bp 'legumin box' sequence, found in the putative promoter regions of a number of legumin genes (see Table 1.2);
3.   sequences homologous to cereal glutenin gene control sequences, which are situated distal to the translation start ATG.

This 5' region flanking of the legumin gene contains 1203 bp upstream of the transcription start site. A series of deletions were made that reduced the length of this putative promoter; the deletions are identified by the number of bases they contain (Figure 9.5c).

–1203	contains all putative control genes
–833	has one of the 'glutenin-control' sequences deleted
–549	has all three of the 'glutenin-control' sequences deleted
–97	has the legumin box plus all three 'glutenin-control' sequences deleted

Transgenic tobacco plants were produced containing one of this deletion series of promoter plus the legumin gene. Legumin gene expression in transgenic tobacco seeds was measured by quantifying the production of the pea protein, using a legumin-specific antibody. The results are shown in Figure 9.5d where levels of legumin are given for a number of transgenic plants containing each promoter construct.

There is considerable variation in the amount of legumin produced in the seeds of the transgenic tobacco plants containing the same construct and this is consistent with the results of transgene expression discussed in section 7.4. Despite this variation a sufficient number of plants were analysed for some conclusions about the promoter to be made. The minimal –97 bp promoter sequence, which only contains the TATA and the CAAT boxes, is not sufficient to provide seed expression of the legumin gene. Increasing the promoter sequence to –549 bp includes the legumin box and this confers a low level of seed expression. However high levels of expression were only seen in plants that contained the legumin transgene plus over 549 bp of 5' flanking sequence, including at least some of the sequences found in cereal glutenin gene promoters. This effect of increasing the level of gene expresion, without affecting the tissue specificity or temporal control of expression, is suggestive of an enhancer sequence (see section 1.5). However, to confirm this, further experiments need to be carried out where the possible enhancer sequences are included in the promoter construct either in a different position relative to the transcription start site or in a different orientation.

This analysis of transgene expression is beginning to provide information about the *cis*-acting regulatory sequences important in determining the extra-

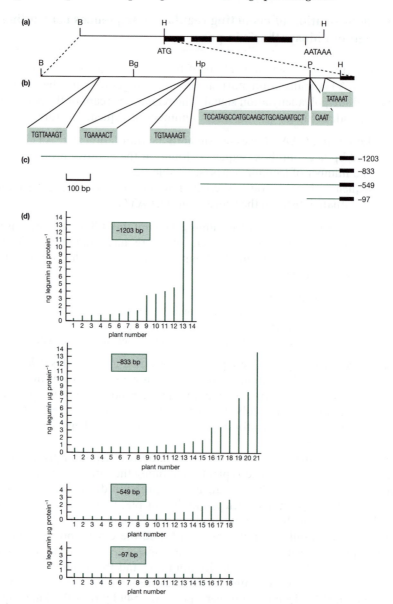

Figure 9.5 Analysis of a pea legumin gene (*leg*A) promoter in transgenic tobacco plants. (a) Map of the *leg*A gene showing exons in solid bars, ATG translation start and polyadenylation signal, AATAAA; (b) expanded 5' flanking region showing TATA box, CAAT box, legumin box and three 'glutenin control' sequences; (c) four promoter regions analysed in transgene expression; (d) production of legumin in seeds of transgenic tobacco plants transformed with each of the four promoters shown in (c). (After Shirsat, A.H., Wilford, N.W., Croy, R.R.D. and Boulter, D. (1989) Sequences responsible for the tissue specific promoter activity of pea legume gene in tobacco. *Molecular and General Genetics*, **215**, 326–331.)

ordinarily high levels of tissue-specific expression of the seed storage protein genes, and complements the analysis of *trans*-acting regulators illustrated by the *opaque*-2 locus in zein synthesis in maize.

Major Learning Objectives for Chapter 9

1. Knowledge of the complexity of seed storage protein structure, with reference to maize zeins and pea legumins and vicilins.
2. Understand the analysis of transcription factors and promoters controlling seed storage protein synthesis.
3. Be acquainted with studies of the intracellular localization of seed storage proteins and appreciate the differences between pea globulin and maize prolamin deposition.
4. Give reasons why the modification of the individual amino acid content of seed storage proteins is an unrealistic strategy for existing transformation technologies.

Further reading

BEWLEY, J.D. and GREENWOOD, J.S. (1990). Protein storage and utilization in seeds. In Dennis, D.T. and Turpin, D.H. (eds) *Plant physiology, biochemistry and molecular biology*, pp. 456–469. Longman, Harlow. This chapter in a multi-author textbook covers the synthesis of seed storage proteins and the formation of protein bodies in a range of species; it also covers mobilization of these seed reserves during germination.

CASEY, R., DOMONEY, C. and SMITH, A.M. (1993). Biochemistry and molecular biology of seed products. In Casey, R. and Davies, D.R. (eds) *Peas: genetics, molecular biology and biotechnology*. pp. 121–164. CAB International, Wallingford.
A clear review of the molecular biology of legumins and vicilins in pea.

LARKINS, B.A., LENDING, C.R. and DE BARROS, E. (1991). Assembly of maize storage proteins into protein bodies in developing endosperm. In Hermann, R.G. and Larkins B.A. (eds) *Plant molecular biology*, vol. 2, pp. 619–625. Plenum Press, New York.
A review of zein synthesis with emphasis on the intracellular localization of zein deposition.

MOTTO, M., DIFONZO, N., HARTINGS, H., MADDALONI, M., SALAMINI, F., SOAVE, C. and THOMPSON, R.D. (1989). Regulatory genes affecting maize storage protein synthesis. *Oxford Surveys of Plant Molecular and Cell Biology*, **6**, 87–114.
A review of the structure and molecular genetics of zein proteins, with emphasis on the mutations which control zein synthesis.

SHEWRY, P.R. and TATHAM, A.S. (1990). The prolamin storage proteins of cereal seeds: structure and evolution. *Biochemical Journal*, **267**, 1–12.
This article contains a great deal of comparative information about the structure of prolamins from maize (zeins) and related species as well as other cereal crops.

Chapter 10

Effect of light on plant development

10.1 Effect of light on plant development

Light controls the normal development of a plant, independently of photosynthesis. One of the most striking illustrations of this is seen in the development of germinating dicotyledon seedlings. Figure 10.1 shows the development of two mustard seedlings of the same age. One, in the light, has a short hypocotyl, expanded cotyledons and photosynthetically active chloroplasts; the other, in the dark, has an elongated hypocotyl and non-photosynthetic cotyledons. The light-grown seedling has undergone photomorphogenesis, whereas the dark developmental programme is called skotomorphogenesis. Skotomorphogenesis occurs because those genes controlling normal plant development are not expressed in the absence of light.

Photomorphogenesis in plants is regulated by at least four different types of photoreceptor:

1. phytochromes, which sense red/far-red light;
2. blue light receptors;
3. ultraviolet (UV)-A absorbing photoreceptor(s);
4. UV-B absorbing photoreceptor(s).

The most thoroughly studied and the best understood of these are the phytochromes and this chapter will be limited to the control of plant development by the phytochrome photoreceptors.

10.2 Criteria for identifying a phytochrome-controlled response

Phytochrome controls seed germination, leaf and stem growth, plastid development and flowering in higher plants. It is a blue protein, is a ubiquitous molecule in higher plants and absorbs red (660 nm) and far-red (730 nm) light.

Figure 10.2 illustrates a method of establishing that phytochrome is involved in a plant response. Flowering plants can broadly be classified as short-day plants (e.g. tobacco) that flower in short days, long-day plants (e.g. barley) that flower in long days and day-neutral plants (e.g. tomato)

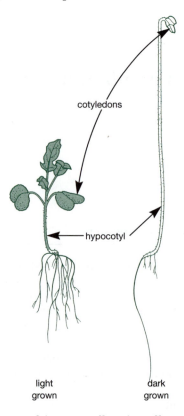

cotyledons

hypocotyl

light
grown

dark
grown

Figure 10.1 Two mustard (*Sinapis alba* L.) seedlings of the same chronological age. The morphological differences are due to the effect of light on the pattern of seedling development. In mustard, germination is epigeal because the cotyledons (as well as having a storage function) appear above ground and differentiate into leaves. In hypogeal germination, found in peas and beans, the cotyledons remain below ground and only have a storage function.

which are insensitive to day length. In short-day and long-day plants it is the length of the dark period (night) that is important and in each species a critical night length can be determined. Figure 10.2 shows the effect of breaking this dark period with a single flash of red light on the flowering response of short- and long-day plants. Thus a single flash of red light in a night longer than the critical length will cause a long-day plant to flower while a short-day plant will remain vegetative. A single flash of far-red light on the other hand has no effect. However, the red light effect is reversible by far-red light, so that giving one red flash followed by one far-red flash leaves the short-day plant flowering. Similarly the far-red effect is reversible by red light so that giving three flashes in the sequence red, far-red, red results in vegetative growth of the short-day plant.

It is the reversible effect of red and far-red light that characterizes a response controlled by phytochrome.

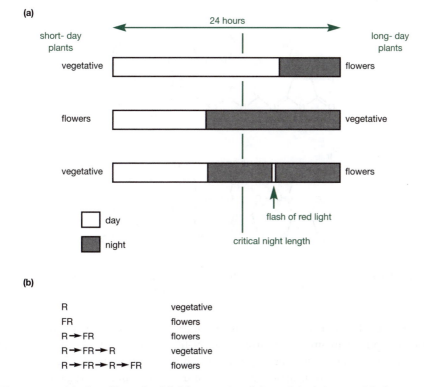

Figure 10.2 (a) The effect of red (R) (660 nm) and far-red (FR) (730 nm) light on flower development in short-day and long-day plants. The light treatment is given as a short flash during a night (dark period) longer than the critical night length. (b) Reversal of light flash response in short-day plants.

10.3 Biochemistry of phytochrome proteins

Phytochromes are cytosol proteins that exist as dimers composed of two $125 \times 10^3\ M_r$ polypeptides. They are not abundant molecules in plant cells and have been difficult to extract and study directly from the plant. Each polypeptide is covalently linked to a tetrapyrrole chromophore (Figure 10.3) in the N-terminal domain. Dimerization occurs through interactions of the C-terminal domains. The photosensory function of the protein is based upon its ability to undergo a reversible interconversion such that the P_r form of the protein absorbs red (R) light and is converted into the P_{fr} form. The P_{fr} form absorbs far-red (FR) light and is converted back to the P_r form. Conformational changes occur in both the polypeptide and in the chromophore (Figure 10.3) during these interconversions.

$$P_r \underset{FR}{\overset{R}{\rightleftharpoons}} P_{fr}$$

Several phytochrome genes have been cloned and studies of the protein have progressed rapidly due to the ability to carry out site-directed mutagenesis and

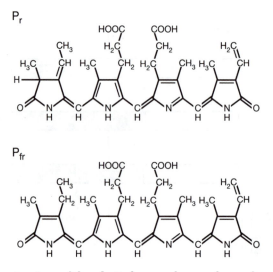

Figure 10.3 The structure of the phytochrome chromophore, showing the difference between the P_r and P_{fr} forms of phytochrome.

reintroduce the modified coding sequence into plants. Figure 10.4 shows a generalized phytochrome structure. The N-terminal region, containing the chromophore, is known to be important in photosensory function. The C-terminal region is involved in dimerization and a domain within this region has been shown to be involved in the regulatory function, that is in the transduction of the light signal to another cellular component. It is proposed that an unknown intramolecular signal transmits the photosignal from the N-terminal domain to the regulatory region. Heterotrimeric G-proteins have been identified as the most likely upstream component of the phytochrome signal transduction pathway (see section 13.1 for further details of signal transduction pathways). Downstream of G-protein activation, biochemical studies indicate that Ca^{2+} and

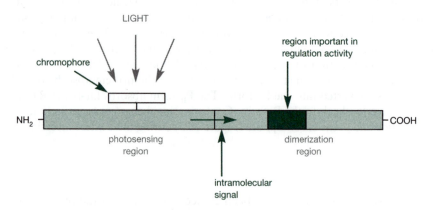

Figure 10.4 Domain structure of phytochrome.

cGMP are involved in phototransduction, possibly in alternative pathways that lead to different cellular responses to red/far-red light signals.

Figure 10.5 gives an overview of the role of phytochrome in controlling plant development. The protein (phytochrome) is a sensory receptor for light (1) which can interpret the light signal (wavelength, intensity) (2). It also has a regulatory function whereby it transmits (transduces) this information (signal) downstream (3) to other components in the cell, which leads to light-responsive gene expression (4). Not all of the steps in Figure 10.5 are understood.

The control of gene expression by phytochrome has been studied using three approaches: (i) characterization of phytochromes themselves, (ii) study of mutants, and (iii) selection of developmental or gene response and tracing the molecular events underlying the response. However, it is the study of mutants that has been the most powerful of these approaches.

10.4 Study of photomorphogenic mutants in *Arabidopsis*

Phytochrome A and B mutants

Most of the studies of photomorphogenesis have been carried out using the etiolation/de-etiolation growth pattern of germinating seedlings seen in Figure 10.1.

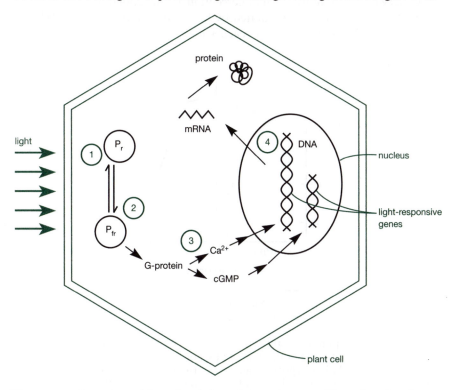

Figure 10.5 Overview of the role of phytochrome in controlling plant development. See text for details.

There are five genes for phytochrome proteins known in *Arabidopsis* and each of the proteins appears to have a distinct function:

*phy*A ⎫ play complementary roles
*phy*B ⎭ in germination
*phy*C ⎫
*phy*D ⎬ role not known
*phy*E ⎭

Genes *phy*A and *phy*B are the best studied and their distinct photosensory functions are illustrated in Figure 10.6, which shows hypocotyl elongation (eti-olated response) in wild type and the *phy*A and *phy*B mutants of *Arabidopsis*. The *phy*A mutant produces no phytochrome A protein (PHYA) but in contin-uous red light (R_c) increasing fluence (amounts) reduces hypocotyl elongation by the same amount as the wild-type plant. However growth of the hypocotyl in the *phy*B mutant (which produces no PHYB protein) is not reduced by increasing fluence of red light. The opposite response is seen for the *phy*A and *phy*B mutants in continuous far-red light (FR_c). Thus the *phy*B mutant can perceive the FR_c light and hypocotyl elongation is reduced, whereas growth of the hypocotyl in *phy*A is not reduced by FR_c. This test indicates that PHYA is necessary for continuous far-red perception and de-etiolation whereas PHYB is necessary for continuous red light perception and de-etiolation.

PHYA synthesis is down-regulated by light, and the P_{fr} form of PHYA is unstable in light; however, PHYA reaches high concentrations in dark-grown

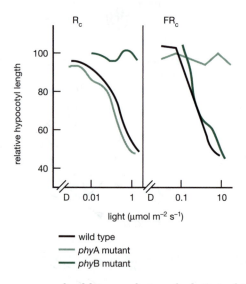

Figure 10.6 The response of wild-type, *phy*A and *phy*B *Arabidopsis* mutants to increasing fluence rates of continuous red (R_c) or far-red (FR_c) light. Response was measured as relative hypocotyl lengths normalized to the dark-grown control for each genotype. D, dark. (Adapted from Quail, P.H., Boylan, M.T., Parkes, B.M., Short, T.W., Xu, Y. and Wagner, D. (1995) Phytochromes: photosensory perception and signal transduction. *Science*, **268**, 675–680.)

seedlings. PHYB, on the other hand, is produced constitutively and the P_{fr} form of PHYB is light stable. In dark-grown seedlings PHYB is the minor form because PHYA is more abundant but in light-grown seedlings PHYB is more important than PHYA.

The antagonistic action of PHYA and PHYB is shown diagrammatically in Figure 10.7, where the solid arrows show the effects of continuous red and far-red light on de-etiolation via PHYB and PHYA respectively. Because $PHYA_{fr}$ is light labile and the synthesis of PHYA is down-regulated by light it is the effect of PHYB that dominates in fully green plants. Open sunlight is red rich so $PHYB_r$ receives red light and is converted to $PHYB_{fr}$, which brings about de-etiolation. However, vegetation shade is less red rich and therefore the conversion of $PHYB_r$ to $PHYB_{fr}$ is reduced and hypocotyls will elongate more than in full sunlight. This is a shade-avoidance response. However PHYA is important in dark-grown seedlings and it may play a critical role in controlling photomorphogenesis during the initial emergence from the soil.

Mutants of the chromophore biosynthesis pathway (*hy*1, *hy*2, *hy*6) have also been identified in *Arabidopsis*. In contrast to *phy*A and *phy*B mutants, which lack a single class of phytochrome, these mutants lack functional molecules of all the phytochromes.

Transduction pathway and regulatory mutants

Figure 10.8 shows the probable point at which each of 13 *Arabidopsis* photomorphogenesis genes acts in the red/far-red light signal transduction pathway.

Mutations of the genes *phy*A, *phy*B, *hy*1, *hy*2 and *hy*6 affect the production of the photoreceptors PHYA and PHYB. The other mutations affect the signal

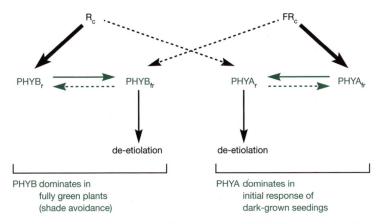

Figure 10.7 Antagonistic action of phytochrome A (PHYA) and phytochrome B (PHYB). Solid arrows indicate the major interaction between each phytochrome and either continuous red (R_c) or far-red (FR_c) light. $PHYA_r$, red light-absorbing form of phytochrome A; $PHYA_{fr}$, far-red light-absorbing form of phytochrome A; $PHYB_r$, red light-absorbing form of phytochrome B; $PHYB_{fr}$, far-red light-absorbing form of phytochrome B.

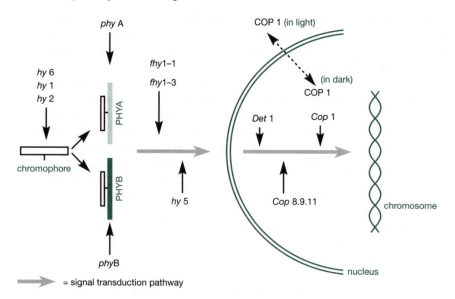

Figure 10.8 The point at which 13 photomorphogenic mutations in *Arabidopsis* are thought to have an effect on red/far-red light reception and signal transduction. See text for details. PHYA, phytochrome A apoprotein; PHYB, phytochrome B apoprotein.

transduction pathway. Mutants of *fhy*1-1, *fhy*1-3 and *hy*5 exhibit etiolated development (skotomorphogenesis) in the dark but have reduced sensitivity to light. That is, they also show etiolated development in the light. Since they appear to have normal levels of phytochrome they are thought to be mutations of genes involved in the signal transduction pathway. The mutants *fhy*1-1 and *fhy*1-3 are only insensitive to far-red light, whereas *hy*5 is insensitive to both red and far-red light.

The other group of mutations, *Cop* and *Det*, are interesting because they lead to seedling development that is de-etiolated in complete darkness. Because the mutations are recessive and their effects are pleiotropic (affect a number of characters), it is deduced that the wild-type products of these genes act to repress the de-etiolated development of seedlings in the dark. That is, they are responsible for dark-induced repression of light-responsive genes.

*Cop*1 is the most intensively studied of this group of genes. It has been cloned and the deduced amino acid sequence suggests that it has a DNA-binding domain (N-terminal zinc-binding motif), together with features found in GTP-binding proteins suggesting protein–protein interaction. This structure is consistent with a role in the repression of light-responsive gene transcription. Transgenic *Arabidopsis* plants that over-express the COP1 protein have reduced sensitivity to light and this is consistent with a repressor molecular function. Thus if COP1 represses de-etiolated growth in the dark, its over-expression could produce repression of de-etiolation in the light.

A mechanism for the light control of *Cop*1 gene repression is suggested by studies of expression of a fusion protein in onion cells. The fusion protein construct consists of the reporter protein GUS fused to COP1. This allows the location of COP1 to be monitored by histochemical staining for GUS. The results are remarkable and show that light affects the intracellular location of COP1 such that in the dark COP1 is located in the nucleus whereas in the light it moves into the cytoplasm.

10.5 Control of gene expression by light

Phytochromes are known to control the transcription of a number of genes, including the small subunit of Rubisco, the chlorophyll *a/b*-binding protein and phenylalanine ammonia lyase. In red light the transcription rate of small subunit (ss) of Rubisco, for example, rises 20-fold.

Figure 10.9 shows the results of an experiment using a series of transgenic tobacco plants. Figure 10.9a shows the DNA construct tested in each transgenic plant and Figure 10.9b shows the expression of the gene in each construct. Two coding sequences are used: the GUS reporter gene and a pea

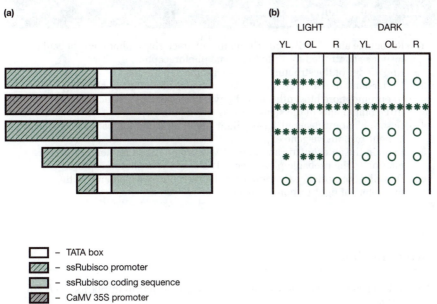

Figure 10.9 The control of pea small subunit Rubisco (ssRubisco) expression by light in transgenic tobacco plants. (a) Five promoter/reporter constructs used to transform tobacco. The TATA box marks the boundary between the promoter sequences tested and the coding sequence of the two reporter genes (ssRubisco and β-glucuronidase, GUS). (b) Expression of each coding sequence in various parts of the transgenic tobacco plants in the light and dark. O, no expression; ***, high levels of expression; *, low level of expression. OL, old leaves; R, roots; YL, young leaves. (Adapted from Moses, P.B. and Chua, N-H. (1988) Light switches for plant genes. *Scientific American*, **258**, 64–69.)

ssRubisco, which can be detected independently from the endogenous tobacco ssRubisco. The promoters tested are either the constitutive CaMV 35S promoter or parts of the pea ssRubisco promoter.

The expression results in Figure 10.9b show that the CaMV 35S promoter drives transcription in all parts of the plant in both the light and dark, whereas the ssRubisco promoter only drives transcription in leaf tissue in the light, that is it is light responsive. Deletions of the ssRubisco promoter reduce the level of light-responsive expression but the light response is not an 'all or nothing' effect of one region of the promoter. Using this type of experiment short DNA sequences have been identified that are important in the control of transcription by light. These DNA sequences are called light response elements (LREs) (see Table 1.2). Light-responsive genes commonly have more than one LRE and this explains the gradual reduction in light responsiveness of deletion series. Table 1.3 lists two protein transcription factors that bind to LRE sequences. Thus the light signal transduction pathway is beginning to be followed from the gene that is light responsive, as well as from the protein (phytochrome) receiving the light signal.

Major Learning Objectives for Chapter 10

1. Recognize and identify the criteria that characterize a photomorphogenic response controlled by a phytochrome photoreceptor.
2. Knowledge of the structure of phytochrome.
3. Understand the distinguishing characteristics of mutations affecting (i) chromophore biosynthesis, (ii) phytochrome protein structure and (iii) components of the signal transduction pathway.
4. Describe in outline a red/far-red light signal transduction pathway that leads to light-responsive gene expression.
5. Explain the antagonistic action of phytochrome A and phytochrome B in relation to the probable physiological roles of these photoreceptors.
6. Describe the evidence that identifies light-responsive elements in the promoters of light-responsive genes.

Further reading

DENG X-W. (1994). Fresh view of light signal transduction in plants. *Cell*, **76**, 423–426.
    A short review of phytochromes, blue and UV-light receptors, together with downstream light-signalling components.

ELICH, T.D. and CHORY, J. (1994). Initial events in phytochrome signalling; still in the dark. *Plant Molecular Biology*, **26**, 1315–1327.
    This is a comprehensive review that gives information about phytochrome structure, the use of mutants and transgenic plants for signal transduction analysis as well as more biochemical studies.

QUAIL, P.H., BOYLAN, M.T., PARKES, B.M., SHORT, T.W., XU, Y. and WAGNER, D. (1995). Phytochromes: photosensory perception and signal transduction. *Science*, **268**, 675–680.
A clear well-illustrated review covering the subject of this chapter.

WHITELAM, G.C. and HARBERD, N.P. (1994). Action and function of phytochrome family members revealed through the study of mutant and transgenic plants. *Plant Cell and Environment*, **17**, 615–625.
This review focuses on the analysis of photomorphogenic mutants of *Arabidopsis* and tomato as well as the transgenic over-expression approach to the study of phytochrome function.

DUCKETT, C.M. and GRAY, J.C. (1995). Illuminating plant development. *Bioessays*, **17**, 101–103.

TERZAGHI, W.B. and CASHMORE, A.R. (1995). Seeing the light in plant development. *Current Biology*, **5**, 466–468.

These last two short articles focus on the analysis of the *Cop* and *Det* mutations of *Arabidopsis*.

Chapter 11

Flowering

11.1 Higher plant (angiosperm) sexual reproduction

Sexual reproduction in higher plants occurs in flowers, and the structure of a schematic flower is shown in Figure 11.1. Floral organ systems develop in four whorls and these are shown in Figure 11.1 as the sepals, petals, stamens and carpels. Pollen is produced in the anthers, which are part of the stamens. Mitotic division of the haploid pollen nucleus gives rise to the male gametophyte generation, which in turn produces the male gamete. The female gamete,

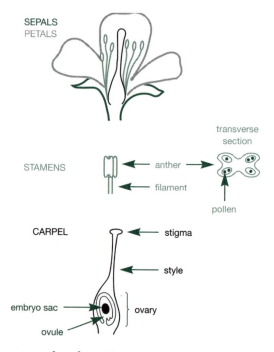

Figure 11.1 Structure of a schematic angiosperm flower showing the four whorls of floral organs: sepals, petals, stamens and carpel.

or egg, is produced by the female gametophyte generation (the embryo sac), which lies within an ovule of the carpel (see section 2.1).

One of the most significant features of the onset of sexual reproduction and therefore flower development is the switch of the plant to a determinate pattern of growth. This begins with the reorganization of an indeterminate vegetative shoot meristem to a determinate flower meristem. In other words, the organ systems of the flower are derived from cells originally present within the vegetative shoot meristem.

Box 11.1 Vegetative versus floral meristem

Cells of the vegetative root and shoot meristems are analogous to animal stem cells because they give rise to new meristematic cells upon division. By contrast, floral meristematic cells differentiate during flower development, that is they become primordia specified to give rise to particular floral organ systems and do not regenerate the meristematic cell line.

There is a progressive and irreversible commitment of the floral primordia to develop into specific organ systems. In the early stages, the floral meristem is able to give rise to all floral organs but as development proceeds individual groups of cells or primordia are committed to differentiate into specific organs.

Tissues within flower organs can be traced by cell lineage studies to particular regions of the floral meristem but development of flower tissues depends not really on strict cell lineage but rather on the position of a cell, or group of cells, in the meristem.

11.2 Genes involved in the regulation of flower development in *Arabidopsis* and *Antirrhinum*

Table 11.1 gives details of a range of *Arabidopsis* mutations affecting flower development. These have been grouped into four classes:

1. genes regulating the change from vegetative to reproductive growth;
2. genes involved in the formation of inflorescence meristems;
3. genes regulating the formation and identity of floral organs;
4. genes regulating the growth of floral organs and their maturation.

In addition, the *Antirrhinum majus* (snapdragon) homologues for a number of the *Arabidopsis* genes have been identified and are included in Table 11.1.

Some of the *Arabidopsis* genes have been cloned and analysis of the deduced amino acid sequence allows a molecular function to be proposed for a proportion of these. The cloned genes have also been investigated by *in situ* hybridization of mRNA. This technique provides information about the spatial and temporal pattern of expression of the genes. In addition, an examination of the mutant phenotype, both of single mutations and of plants

Table 11.1 Genes involved in the regulation of flower development

Arabidopsis gene	Mutant phenotype	Proposed molecular function	*Antirrhinum* homologue
Genes regulating vegetative to reproductive growth			
*ga*1	Late flowering	Defective in gibberellin biosynthesis, *ent*-kaurene synthase	ND
*phy*A	Long hypocotyl in continuous far-red light; less sensitive to night break	Phytochrome A apoprotein	ND
*hy*3 (*phy*B)	Early flowering, long hypocotyl	Phytochrome B apoprotein	ND
*hy*2	Early flowering, long hypocotyl	Phytochrome chromophore biosynthesis	ND
fca	Late flowering, vernalization responsive	ND	ND
Genes involved in formation of inflorescence and floral meristems			
tfl (*Terminal flower*)	Conversion of inflorescence meristem to floral meristem	Negative regulator of *lfy*, *ap*1, *ap*2	ND
lfy (*Leafy*)	Partial conversion of inflorescence meristem to floral meristems	Transcription factor	*flo* (*Floricaula*)
*ap*1 (*Apetala*1)	Production of auxiliary flowers in flowers; homeotic conversion of sepals to leaves	MADS box transcription factor	*squa* (*Squamosa*)
*ap*2 (*Apetala*2)	Homeotic conversion of sepals to leaves and petals to stamens in weak mutants; homeotic conversion of sepals to carpels/petals and stamens absent in strong mutants	Negative regulator of *Agamous* (*ag*), homology to DNA-binding domain of ethylene-responsive element binding proteins (see section 13.2)	ND
pin (*Pin formed*)	Forming no floral buds or deformed flowers	A component of auxin polar transport system	ND
Genes regulating formation and identity determination of floral organs			
*ap*1	See above		
*ap*2	See above		
*ap*3 (*Apetala*3)	Homeotic conversion of petals to sepals and stamens to carpels	MADS box transcription factor	*def*(*Deficiens*)
ag (*Agamous*)	Homeotic conversion of stamens to petals; indeterminate floral meristem; no pistil	MADS box transcription factor	*ple*(*Plena*)
pi (*Pistillata*)	Homeotic conversion of petals to sepals and stamens to carpels	MADS box transcription factor	*glo* (*Globosa*)

Table 11.1 (continued)

Arabidopsis gene	Mutant phenotype	Proposed molecular function	Antirrhinum homologue
sup (*Superman*)	More stamens; small pistil	Regulator of *pi* and *ap*3 expression	ND
Genes regulating floral organ growth and maturation			
at (*Antherless*)	Stamens lacking anthers	ND	ND
apt	Abortive pollen development after meiosis	Adenine phosphoribosyl-transferase	ND
*sin*1 (*Short integuments*)	Abortive ovules	ND	ND

ND, not determined.

containing more than one mutation, gives information about the interaction of these genes. Together these studies are beginning to produce a picture of the genetic control of flower development in dicotyledons.

In this chapter a selection of the genes shown in Table 11.1 will be discussed in order to illustrate the important controls of flower development that have recently been identified.

Class 1: genes regulating the change from vegetative to reproductive growth

The genes in this class are of three types: they are involved in gibberellin (a plant hormone) biosynthesis, they are phytochrome genes or they are vernalization genes. Gibberellic acid has been implicated in the homeostasis of flower development under different environments (see section 11.3) but its precise biochemical role is not known. The role of phytochromes in the control of flowering by day length has been covered in Chapter 10.

Late flowering (*fca*)

In *Arabidopsis* the transition from vegetative to reproductive development (floral induction) results in the end of rosette leaf formation and the rapid appearance of the inflorescence with cauline leaves and flowers (Figure 11.2). Floral induction in some ecotypes is accelerated by a low-temperature treatment (4°C for 4–8 weeks), which is termed vernalization. Table 11.2 shows the effect of the *fca* mutation on flowering of the early flowering 'wild-type' ecotype of *Arabidopsis* Landsberg *erecta*. The *fca* homozygous mutants are late flowering but an 8-week vernalization treatment of sown seeds produces a flowering development similar to the 'wild-type' plants. This gene is now being cloned using linked molecular markers and a chromosome walking technique.

Table 11.2 Effect of vernalization on flowering of *Arabidopsis thaliana*. (Adapted from Chandler, J. and Dean, C. (1994) Factors influencing the vernalisation response and flowering time of late flowering mutants of *Arabidopsis thaliana* (L.) Heynh. *Journal of Experimental Botany*, **45**, 1279–1288)

Genotype	Leaf number (a)			Flowering time (b)		
	No vernalization	Vernalization	Per cent reduction	No vernalization	Vernalization	Per cent reduction
Landsberg *erecta* (wild type)	9.0 ± 0.2	8.1 ± 0.2	10	22.9 ± 0.5	17.0 ± 0.4	26
fca-1 (late flowering mutant homozygote)	29.0 ± 1.2	8.3 ± 0.3	71	47.1 ± 2.5	16.3 ± 0.2	65

(a) At flowering.
(b) Number of days' growth before production of inflorescence.

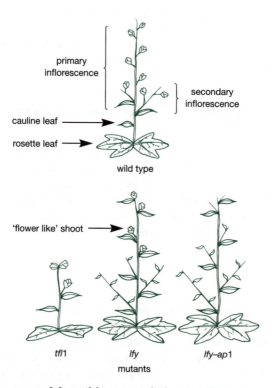

Figure 11.2 Structure of the wild-type *Arabidopsis* inflorescence together with the structure of the inflorescence of homozygous *tfl*1; *lfy* and *lfy–ap*1 (double-mutant) plants. (Adapted from Okamuro, J.K., den Boer, B.G.W. and Jofuku, K.D. (1993) Regulation of *Arabidopsis* flower development. *Plant Cell*, **5**, 1183–1193.)

Class 2: genes involved in the formation of inflorescence meristems

Terminal flower (*TFL1*)

Figure 11.2 shows the structure of a wild-type *Arabidopsis* inflorescence together with the structure of a homozygous *tfl1* mutant inflorescence. Within an inflorescence each flower and inflorescence meristem must maintain a separate identity. The *tfl1* homozygous mutants have lost the maintenance of the inflorescence meristem and flowers are produced prematurely. The *tfl1* gene is therefore thought to have a role in inflorescence meristem maintenance.

Leafy (*LFY*)

Leafy plays a central role in the establishment of the floral meristem and in the regulation of flower gene expression. Mutations of *Leafy* cause a delay in floral meristem production by the inflorescence, and Figure 11.2 shows that homozygous mutants also produce leafy structures instead of flowers. The gene encodes a proline-rich protein which is possibly a novel transcription factor.

The *Leafy* gene is expressed early in the cascade of flower-specific genes in all of the floral primordia, and Figure 11.3 shows this spatial pattern of *Leafy* gene expression during flower development.

The gene *Apetala1* is included in both class 2 and class 3 because it is thought to be involved in both the development of the floral meristem and in the production of determinate primordia within the meristem.

Class 3: genes regulating the formation and identity of floral organs

Homeotic mutations cause the conversion of one organ into another and this distinctive phenotype implicates the genes in the control of organ function. Most of the genes in class 3 have been identified by the production of homeotic mutations.

Apetala1 (*AP1*)

This gene has a dual role: (i) in early expression it is thought to be involved in the conversion of the inflorescence meristem to a floral meristem; (ii) in later expression it is thought to be involved in the formation of sepal primordia. Mutants of *Apetala1* have a partial conversion of the floral meristem to an inflorescence meristem; they also have a homeotic conversion of sepals to leaves. Figure 11.2 shows the interaction of *ap1* and *lfy* mutations, which in the double-mutant plant produce no flowers because all of the floral meristems are indeterminate shoot meristems.

The temporal and spatial pattern of *Apetala1* expression is shown in Figure 11.3. The gene is expressed early in development but unlike *Leafy* it continues to be expressed during the formation of flower organs, where it is restricted to two whorls, the petals and sepals.

developmental
stage

MADS box genes

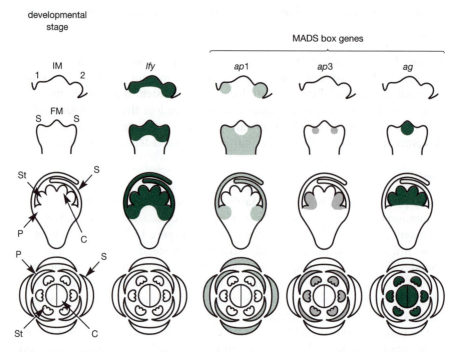

Figure 11.3 Temporal and spatial regulation of *Arabidopsis* meristem and organ identity gene expression during flower development. C, carpels; FM, floral meristem; IM, inflorescence meristem; P, petals; S, sepals; St, stamens. *ag*, *Agamous*; *ap*1, *Apetala*1; *ap*3, *Apetala*3; *lfy*, *Leafy*. (Adapted from Okamura, J.K., den Boer, B.G.W. and Jofuku, K.D. (1993) Regulation of *Arabidopsis* flower development. *Plant Cell*, **5**, 1183–1193.)

Apetala3 (AP3)

Apetala3 is expressed later in flower development than *Apetala1* and it is not active in the inflorescence meristems (Figure 11.3). In the developing flower it is expressed in two whorls, the petals and stamens. Mutations in *Apetala3* cause the homeotic conversion of petals to sepals and stamens to carpels.

Agamous (AG)

Agamous is important in regulating the second half of flower development, that is the development of stamens and carpels and the termination of flower meristematic activity. Mutant *Agamous* plants produce flowers with carpels and have the homeotic conversion of stamens to petals. In addition, the floral meristem is indeterminate, leading to flowers within flowers.

Figure 11.3 shows that *Agamous* is expressed at the same stages of flower development as *Apetala3* but its expression is restricted to two whorls, carpels and stamens.

Analysis of the deduced amino acid sequences of *Apetala1*, *Apetala3* and *Agamous* show that they all contain a highly conserved 58-amino acid

DNA-binding domain, which is called the MADS box and which is found in a conserved family of transcription factors. The acronym MADS comes from yeast *minichromosome maintenance*1, *Arabidopsis* *ag*, *Antirrhinum* *def* and human *SRF*. This homology suggests that these genes encode transcription factors that control the expression of other genes.

Class 4: genes involved in floral organ growth and maturation

Antherless (*at*)

Antherless is an example of the class of gene involved in the growth and maturation of floral organs rather than the initiation of their formation. Plants homozygous for mutations in the *Antherless* gene produce a whorl of stamens but these lack anthers and therefore produce no pollen.

11.3 Homeostasis of flower development

Arabidopsis is a quantitative long-day plant. Changes in photoperiod and in temperature can dramatically alter the timing of flowering but they do not change the pattern of inflorescence and flower development, which are, in wild-type plants, resistant to environmental changes. This is termed homeostasis.

Each of the five mutants described in classes 2 and 3 have a classic long-day 'signature' phenotype. However, each mutant can have a changed phenotype when the mutant plant is grown in an altered photoperiod or temperature. This suggests that the normal alleles of these genes are involved in maintaining the normal pattern of development under different environmental conditions and this is lost in the mutant. For example, a short-day treatment of *agamous* mutant plants induces a reversion from floral meristems to inflorescence meristems.

Gibberellic acid is implicated in this homeostasis because applications of gibberellin to mutants grown in short days restore the long-day phenotype.

11.4 Temporal regulation of gene expression during flower development

Figure 11.4 shows a model of the temporal regulatory network of flower development. This network proposes that there is a promotive interaction between genes in the different classes shown in Table 11.1. Thus expression of the genes in class 2 leads, by an unknown mechanism, to the expression of genes in class 3. However, this promotive interaction of some genes must be modified by additional antagonistic interactions. This is illustrated in Figure 11.4 by the interaction of *Terminal flower* and other class 2 genes. Loss of function of the *Terminal flower* gene leads to the loss of suppression of the genes for floral meristem formation and *Terminal flower* mutants produce flowers prematurely on the inflorescence (Figure 11.2).

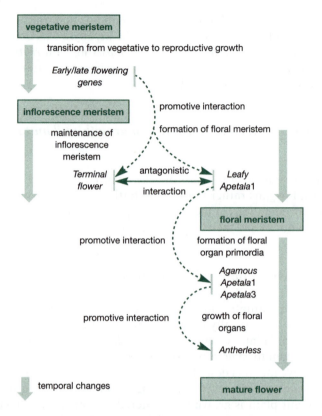

Figure 11.4 Temporal regulatory network of flower development showing the stage at which the *Arabidopsis* genes, identified by mutations, act. Genes described in text are shown in italics. (Adapted from Okada, K. and Shimura, Y. (1994) Genetic analysis of signalling in flower development using *Arabidopsis*. *Plant Molecular Biology*, **26**, 1357–1377.)

11.5 Spatial regulation of gene expression during flower development

Figure 11.5 shows a model for the mechanism by which spatial interaction of genes leads to the production of primordia that have different floral organ identities. In this model the four whorls require different gene functions:

whorl 1 (sepals) requires 'a' gene functions
whorl 2 (petals) requires 'a' plus 'b' gene functions
whorl 3 (stamens) requires 'b' plus 'c' gene functions
whorl 4 (carpels) requires 'c' gene function
*Apetala*1 is an example of an 'a' gene
*Apetala*3 is an example of a 'b' gene
Agamous is an example of a 'c' gene

Reference to Figure 11.3 will show that the spatial pattern of expression of these MADS box genes is consistent with this model. Thus *Apetala*1 ('a' gene)

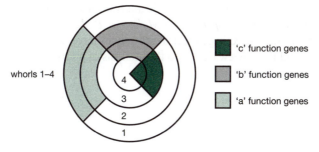

Figure 11.5 The four whorl regions of a flower meristem and the regions of expression of three classes of organ homeotic genes. Whorl 1, sepals; whorl 2, petals; whorl 3, stamens; whorl 4, carpels. (Adapted from Coen, E.S. and Meyerowitz, E.M. (1991) The war of the whorls: genetic interactions controlling flower development. *Nature*, **353**, 31–37.)

is expressed in sepals and petals but *Agamous* ('c' gene) is expressed in stamens and carpels. The expression of *Apetala3* ('b' gene) overlaps the expression of both *Apetala1* and *Agamous* since it is expressed in petals and stamens. The interaction of *Apetala1* and *Agamous* is therefore antagonistic, whereas the interaction of *Apetala3* is synergistic with both *Apetala1* and *Agamous*.

Major Learning Objectives for Chapter 11

1. Knowledge of the key features of a floral meristem which distinguish it from a vegetative meristem.
2. Understand how the pattern of temporal regulation of genes has been derived from analysis of mutations that lead to changes in flower development.
3. Understand how the model for the spatial regulation of genes during flower development has been derived from an analysis of homeotic mutations.
4. Define and explain homeostasis of flower development.

Further reading

The Plant Cell, Special Issue 5, 1139–1488 (1993). This issue contains 28 review articles on plant reproduction; it is very comprehensive in its coverage and contains some well-illustrated and clear reviews that cover in more detail the material of this chapter.

DREWS, G.N. and GOLDBERG, R.B. (1989). Genetic control of flower development. *Trends in Genetics*, **5**, 256–261.
A good introduction to the topic.

COEN, E.S. and MEYEROWITZ, E.M. (1991). The war of the whorls: genetic interactions controlling flower development. *Nature*, **353**, 31–37.
A review of homeotic mutations and floral organ formation in *Arabidopsis* and *Antirrhinum*.

OKADA, K. and SHIMURA, Y. (1994). Genetic analysis of signalling in flower development using *Arabidopsis*. *Plant Molecular Biology*, **26**, 1357–1377. More than 60 genes affecting flower development have been identified in *Arabidopsis* and this review describes the analysis of a large selection of them.

Chapter 12

Breeding systems

12.1 Genetic control of self-incompatibility

In 90 of the 320 higher plant (angiosperm) families, it has been reported that
self-fertilization is blocked by a genetically controlled mechanism. In several
families the control of this block, which is known as self-incompatibility, can
be attributed to a single locus, the S locus. However, this locus may in fact represent a group of closely linked genes.

The S loci have a number of characteristics that are common in all families.

1. There are multiple forms of the S locus; in some species about 200
   different S locus forms are estimated. Because the S locus may consist of
   a group of linked genes, it is possible that in some cases the variation
   between S loci is not allelic. For this reason the term allele is qualified in
   this chapter by the use of quotation marks ('allelic').
2. Any two S 'alleles' may be present in a diploid plant.
3. Incompatibility occurs when the S 'allele(s)' in the male parent is the
   same as the S 'allele(s)' in the female parent.

The S-locus-controlled incompatibility systems of higher plants are mechanisms
by which **self** can be distinguished from **non-self**. They involve intercell signals
and as such provide a challenge to plant molecular geneticists. Rapid advances
in the study of the molecular and biochemical processes of self-incompatibility
are being made and these reveal an unexpected variety of systems.

Figure 12.1 shows the key reproductive tissues involved in incompatibility
mechanisms. Successful fertilization occurs when a pollen grain reaches the
stigma and germinates to produce a pollen tube, which penetrates the transmitting tissue of the style and grows down to the ovules within the ovary. The single
haploid nucleus in the early pollen grain undergoes haploid mitoses to produce
the three haploid nuclei in the germinating pollen. One of these nuclei is present
in a vegetative cell and the other two travel down the elongating pollen tube.

The double fertilization of higher plants (see section 2.1) occurs after the
pollen tube has entered the ovule micropyle, when the male gamete nucleus fuses
with the female egg nucleus to form the diploid zygote (the next sporophyte

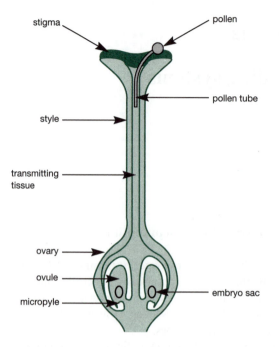

Figure 12.1 Reproductive tissues involved in self-incompatibility mechanisms.

generation) and the second male nucleus fuses with the two polar nuclei to form the triploid endosperm.

12.2 Gametophytic versus sporophytic incompatibility mechanisms

There are two different genetic mechanisms involved in S-locus-controlled self-incompatibility, known as gametophytic incompatibility and sporophytic incompatibility. The key difference between them is that in gametophytic incompatibility it is the genotype of the haploid pollen (gametophyte) that determines the pollen incompatibility reaction, whereas in sporophytic incompatibility it is the diploid genotype of the plant (sporophyte) that produced the pollen which determines the pollen incompatibility reaction. Among higher plant families gametophytic incompatibility is more common than sporophytic incompatibility.

The genetic control of the gametophytic and sporophytic incompatibility mechanisms is illustrated in Figure 12.2. For each pollination the genotype of the pollen parent in Figure 12.2 is the same (S_1S_3) but three different female parents are illustrated: S_1S_2, S_2S_3 and S_2S_4. This means that two of the female parents shown have one S 'allele' in common with the male parent (either S_1 or S_3) and only the S_2S_4 female parent has S-'alleles' not present in the male parent.

With a gametophytic incompatibility system (such as that found in *Nicotiana alata*), the pollen is phenotypically either S_1 or S_3 and this means that S_3 pollen

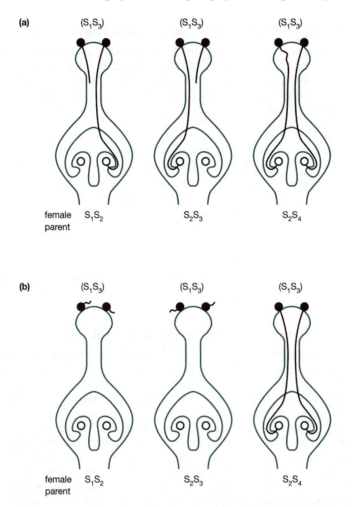

Figure 12.2 Incompatible and compatible crosses involving the pollen parent, genotype S_1S_3, with (a) a gametophytic self-incompatibility system and (b) a sporophytic self-incompatibility system. The same three female genotypes (S_1S_2, S_2S_3 and S_2S_4) are shown in each incompatibility system.

can fertilize the S_1S_2 female and S_1 pollen can fertilize the S_2S_3 female. Both types of pollen can fertilize the S_2S_4 female parent.

With a sporophytic incompatibility system (such as that found in *Brassica oleracea*), however, the pollen is phenotypically S_1S_3 and if either the S_1 'allele' or the S_3 'allele' is present in the female parent, fertilization is prevented. The only successful fertilization will be with the S_2S_4 female parent, which has neither the S_1 nor the S_3 'allele'. In the example illustrated in Figure 12.2 the S 'alleles' are co-dominant. In some species the S 'alleles' can show dominance and in this case all the pollen produced by a heterozygous plant will have the phenotype (incompatibility reaction) of the dominant S 'allele'.

> **Box 12.1 Timing of expression of incompatibility system in pollen**
>
> The difference between the two systems may involve timing of action of the genes controlling incompatible reactions. In the gametophytic system they must be expressed after meiosis in the male haploid gametophyte but in the sporophyte system they must be expressed before male meiosis in the diploid sporophyte cell.

There are other differences between gametophytic and sporophytic systems. In a gametophytic incompatibility reaction the pollen germination and growth of the pollen tube is inhibited within the style tissue. In a sporophytic incompatibility reaction, the pollen tube growth is inhibited on the stigma surface. There are also differences between the two systems in the maturity of the pollen at pollination. With the exception of Gramineae, gametophytic pollen is binucleate whereas sporophytic pollen is trinucleate.

Table 12.1 shows the breeding system for a range of higher plant species. This illustrates that a single family may have both self-fertile and self-incompatible species, but within a family the incompatibility system is the same in all species.

> **Box 12.2 Homomorphic and heteromorphic incompatibility**
>
> In most of the families with an S-locus-controlled incompatibility mechanism the morphology of the flowers is identical in all members of the species. However in 23 families there are genes for floral morphology linked to the S locus. These plants have a heteromorphic incompatibility mechanism and primrose (*Primula vulgaris*) is an example of this type of mechanism. In *Primula*, which has a sporophytic incompatibility system, the number of S 'alleles' is limited to two, and these also control alternate positions of the stigma and anthers within the flowers, such that in pin flowers the stigma is above the anthers and in thrum flowers the anthers are above the stigma. Plants producing pin flowers are genetically *ss*, whereas plants producing thrum flowers are *Ss*. Crosses between pin and thrum plants will therefore produce 50% pin progeny and 50% thrum progeny.

12.3 Molecular studies of sporophytic incompatibility in *Brassica oleracea*

Molecular studies of incompatibility in *Brassica* began with the identification of an abundant class of protein synthesized by the stigma 1 day before anthesis (when pollen is shed from the anthers). These proteins are between 45 and 55×10^3 M_r in size, are heavily glycosylated and have been called S-locus-specific glycoproteins (SLSG). A 2-kb SLSG cDNA clone was isolated by differentially screening a stigma cDNA library. This clone represented an mRNA produced in stigmas at the critical time of SLSG synthesis (just as the flowers are about to open); the mRNA was not produced in either leaves or seedling tissues. The SLSG mRNA was found to be absent from self-compatible *Brassica* mutants.

Table 12.1 Breeding system of some plant species (within a single family the incompatibility system is the same in all self-incompatible species)

		Breeding system	
Family	Species	Self-fertile	Self-incompatible
Cruciferae	*Arabidopsis thaliana*	Yes	
	Brassica napus	Yes	
	Brassica oleracea		Sporophytic incompatibility
Gramineae (two-locus incompatibility mechanism)	*Hordeum vulgare*	Yes	
	Secale cereale		Gametophytic incompatibility
Leguminosae	*Trifolium repens*		Gametophytic incompatibility
	Pisum sativum	Yes	
Papaveraceae	*Papaver rhoeas*		Gametophytic incompatibility
Primulaceae	*Primula vulgaris*		Sporophytic heteromorphic incompatibility
Scrophulariaceae	*Antirrhinum majus*	Yes	
Solanaceae	*Nicotiana alata*		Gametophytic incompatibility
	Nicotiana tabacum	Yes	
	Solanum tuberosum		Gametophytic incompatibility
	Petunia hybrida	Yes	
	Petunia inflata		Gametophytic incompatibility

In situ hybridization showed that the SLSG are present in the cell walls of the stigma cells (known as papillae).

Co-segregation of S 'alleles' and SLSG genomic restriction fragments

Southern blots of *Brassica* genomic DNA digested with a restriction endonuclease and probed with the SLSG cDNA clone (pBOS5) produce a number of restriction fragments that hybridize to the cDNA. Further, the pattern of restriction fragments produced varies between plants (i.e. RFLP; see section 2.2). Figure 12.3 shows the pattern of restriction fragments seen in a genomic Southern blot of DNA extracted from *Brassica* plants possessing different S 'alleles' that have been crossed to produce an F_1 and an F_2 generation.

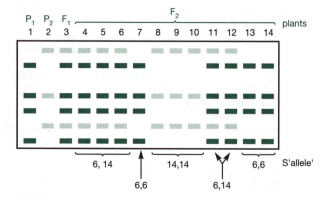

Figure 12.3 Genomic Southern blot of DNA extracted from two parent *Brassica* plants, the F_1 hybrid and a sample of F_2 plants. The Southern blot was probed with the S-locus-specific glycoprotein cDNA clone, pBOS5, and shows RFLP between the parent plants and between F_2 progeny (for details see text). P_1, S_6 S_6; P_2, S_{14} S_{14}; F_1, S_6 S_{14}. (Adapted from Nasrallah, J.B., Kao, T-H., Goldberg, M.L. and Nasrallah, M.E. (1985) A cDNA clone encoding an S-locus-specific glycoprotein from *Brassica oleracea*. *Nature*, **318**, 263–267.)

Box 12.3 Bud pollination

Self-fertilization can be forced in *Brassica* by pollinating immature flowers and is necessary in this experiment both to produce the homozygous S 'allele' parents (S_6S_6 and $S_{14}S_{14}$) and to produce the F_2 generation.

Parent 1 (S_6S_6) shown in Figure 12.3, has four genomic DNA restriction fragments that hybridize with the SLSG cDNA clone whereas parent 2 ($S_{14}S_{14}$) has only three homologous genomic fragments. One restriction fragment is common to both parents and all the progeny but the others vary in size, and therefore mobility in the gel, between parent 1 and parent 2. The F_1 plant (S_6S_{14}) produced by crossing these two parents contains all the fragments found in both parents. When the F_1 hybrid plant is self-fertilized the S 'alleles' will segregate and individual F_2 progeny will inherit either S_6 or S_{14} from the male gamete and either S_6 or S_{14} from the female gamete. This means that the F_2 progeny may be S_6S_6, $S_{14}S_{14}$ or S_6S_{14} depending on the combination of S 'alleles' in the two gametes.

Inspection of the restriction fragment patterns found in the F_2 plants (Figure 12.3) shows that three patterns exist: the four-fragment pattern of parent 1; the three-fragment pattern of parent 2; or the six-fragment pattern of hybrid F_1. Further, each fragment pattern is present in a plant with the equivalent S-locus 'alleles' of parent 1, parent 2 or F_1. In other words, if an F_2 plant is S_6S_6 (like parent 1) it has the parent 1 restriction fragment pattern, whereas an F_2 plant with $S_{14}S_{14}$ has the parent 2 restriction fragment pattern and an S_6S_{14} F_2 plant has the F_1 restriction fragment pattern.

The S 'allele' is said to co-segregate with the restriction fragment pattern and this indicates that the S locus and the gene coding for SLSG are linked. This in turn is strong evidence that SLSG are involved in the incompatibility reaction of S 'alleles'.

Box 12.4 Variation in SLSG between different S 'alleles'

A number of SLSG genes have now been cloned from plants containing different S 'alleles' and it has been shown that there are many amino acid changes between these 'alleles' and that the amino acid variation is clustered to certain regions of the SLSG polypeptide.

S-related kinase

During the process of screening a *Brassica* genomic library for SLSG genes, another gene was isolated because it had some sequence homology with the SLSG cDNA clone used as a probe. The deduced amino acid sequence of the new gene shows three domains: (i) an SLSG-like N-terminal domain, (ii) a transmembrane domain, and (iii) a serine/threonine kinase (protein kinase) domain. This gene, called SRK (S-related kinase), is linked to the S locus and SLSG, and is also expressed in stylar cells. When different plants are compared the results show that there is more sequence homology between the SLSG and the linked SRK N-terminal domains on the same chromosome than between different SLSG alleles.

The isolation of this protein kinase gene and the well-established mechanism of protein phosphorylation (by similar kinases) in animal signal transduction (see Chapter 13) has led to the formulation of a model for the interaction between pollen and stigma papillae in *Brassica*.

Model for interaction of pollen and stigma papillae

Figure 12.4 shows a self-incompatible pollination in *Brassica*, summarizing what is known about the interaction of pollen and a stigma papilla.

An unknown pollen molecule, the ligand, is thought to interact with both the cell wall-located SLSG and the SRK, which is thought to be located in the plasma membrane. This interaction may activate the kinase domain of SRK, which will then phosphorylate another protein. Phosphorylation cascades are involved in signal transduction in animal cells and a similar mechanism may occur in the stigma cells, leading to the production of a response, which in turn inhibits pollen development.

12.4 Molecular studies of gametophytic incompatibility in *Nicotiana* and *Petunia*

Nicotiana spp. have a gametophytic incompatibility mechanism and molecular studies began with the identification of a major stylar glycoprotein. These

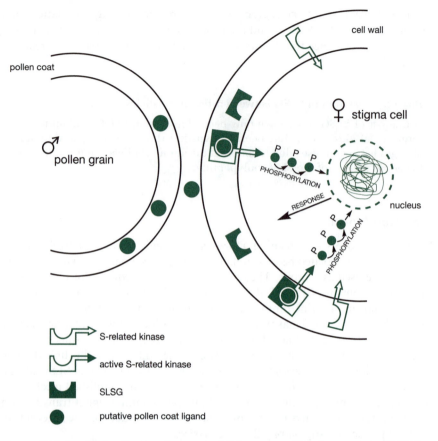

Figure 12.4 Model for the involvement of S-locus-specific glycoprotein (SLSG) and S-related kinase in the incompatible reaction of *Brassica* stigma cells.

proteins are smaller (24–32×10^3 M_r) than the SLSG from *Brassica*. Using a differential (mature versus immature) screen a mature style cDNA clone was isolated; this was confirmed as the gene for the stylar glycoprotein by a comparison of the deduced amino acid sequence of the clone and the N-terminal amino acid sequence of the glycoprotein itself. The mRNA for this gene is found in mature styles and ovaries and the protein is found in large quantities in the intercellular matrix at the top of the style transmitting tract.

Sequence analysis of the cDNA clone shows that it is not related to the *Brassica* SLSG but that it has homology with a group of fungal ribonucleases (RNase) and has consequently been named S-RNase. Table 12.2 shows some circumstantial evidence for the production of an RNA-degrading enzyme during incompatible pollinations in *Nicotiana alata*. Plants of different genotypes were pollinated with pollen grains that had previously had their nucleic acids radio-labelled with ^{32}P. After allowing time for the pollen to germinate the RNA was extracted from the pollinated styles and the amount of radioactive RNA (which comes from the pollen) was calculated. Table 12.2 shows that less pollen RNA is

Table 12.2 Self-incompatibility in *Nicotiana alata* involves degradation of pollen RNA recovered from extracts of pollinated styles

Genotype of style		Genotype of pollen		^{32}P RNA (counts min^{-1})
♀ S_1S_3	×	S_2S_2 ♂	Compatible	161
♀ S_2S_2	×	S_2S_2 ♂	Incompatible	49
♀ S_2S_2	×	S_1S_3 ♂	Compatible	165
♀ S_1S_3	×	S_1S_3 ♂	Incompatible	43

extracted from incompatible crosses, and this supports the proposal that an RNase is produced during incompatible crosses.

More recently, two experiments with transgenic plants have provided confirmation of the role of an S-RNase in incompatibility. These experiments have been carried out in *Petunia inflata*, a species related to *N. alata* (Solanaceae) which also has a gametophytic incompatibility mechanism.

Box 12.5 Incompatibility in polyploids

Expression of transgenes in *N. alata* proved to be difficult and therefore a related species (*P. inflata*), which is easy to transform, was chosen. It is not possible to test the effects of removing S 'allele' function in *N. tabacum* because it is self-compatible. *N. tabacum* is a polyploid species and it is common for polyploid species to be self-compatible. However this is not always the case: *Trifolium repens* (white clover) is an example of a polyploid species with a gametophytic incompatibility system.

Two types of transgenic *P. inflata* were produced. In the first experiment *Petunia* plants were transformed with the antisense construct of an S-RNase controlled by the style-specific S-RNase promoter. This led to the inhibition of S-RNase activity in the styles of the transgenic plants, which were also found to be self-compatible. In this experiment the S-RNase clone (S_3) used for the antisense construct came from the plant (S_2S_3) that was transformed. In the second experiment a sense S_3 S-RNase construct was used to transform S_1S_2 *Petunia* plants. In this case the transgenic plants have become $S_1S_2S_3$ and were found to reject S_3 pollen as well as S_1 and S_2 pollen. The incompatibility reaction of the pollen was not affected by the transgene in either experiment, that is the pollen was either S_1 or S_2. These two experiments provide evidence that the S-RNase proteins control the self-incompatibility behaviour of the style in *Petunia*.

Model for interaction of pollen and style

Figure 12.5 outlines a scheme for the interaction of pollen and style transmitting tract cells that incorporates S-RNase activity. Inactive S-RNase is

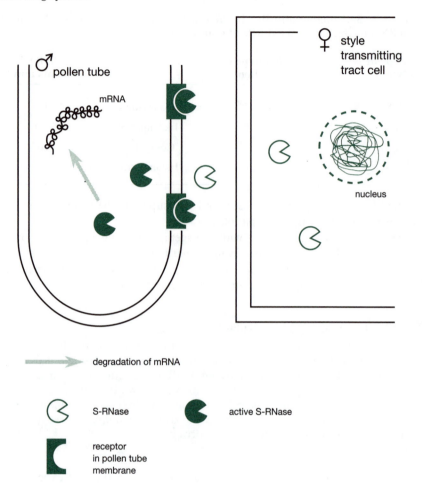

Figure 12.5 Model for the role of style transmitting tract S-RNase in the incompatible reaction of pollen tubes in *Nicotiana* and *Petunia*.

produced by the style cells, and this protein is activated and transported into the pollen tube by a putative pollen tube receptor. Within the growing pollen tube the active S-RNase degrades RNA leading to a failure of growth of the tube and a failure of fertilization.

12.5 Molecular studies of gametophytic incompatibility in *Papaver rhoeas* (poppy)

Species of poppy have a gametophytic incompatibility mechanism and the gene for a small stigma protein has been cloned. This protein is not an S-RNase nor an SLSG but like these proteins it shows complete linkage with the poppy S locus. The protein has been expressed in *E. coli* and the recombinant protein has been shown to inhibit pollen grain germination *in vitro*.

Box 12.6 Poppy flower structure

The inhibition of pollen development in the gametophytic incompatibility system of poppy occurs on the stigma surface, unlike other gametophytic systems where inhibition occurs in the style. This is because there is no style in the poppy pistil.

It has been shown that the self-incompatible response in poppy is mediated by the release of Ca^{2+} into the pollen cytoplasm and this has led to the model for pollen–stigma interaction illustrated in Figure 12.6.

Model for interaction of pollen and stigma

The small stigma protein is released from the stigma cells. It is not internalized into the pollen cell but is presumably recognized by a receptor on the

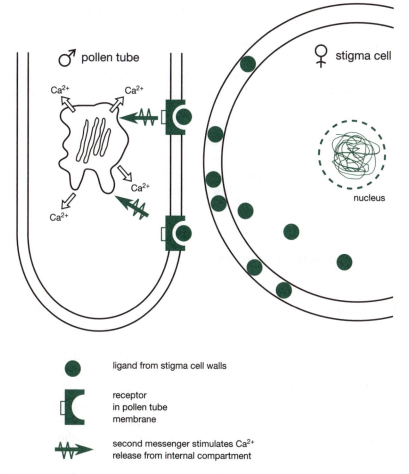

Figure 12.6 Model for the role of a small stigma protein in the inhibition of pollen tube growth in an incompatible cross in *Papaver*.

pollen tube surface. It is thought that the binding of the receptor and small stigma protein generates a second message in the pollen tube, which in turn stimulates the release of Ca^{2+} from an intracellular compartment into the cytoplasm. The resulting increase in cytoplasmic levels of Ca^{2+} inhibits pollen tube growth.

12.6 Unanswered questions in self-incompatibility

All the information in sections 12.3, 12.4 and 12.5 relates to the basis of S-locus activity in the female flower organs (stigma and style).

Although the biochemical basis of incompatibility appears to be different in *Brassica*, *Nicotiana* and *Papaver*, in each system the incompatible reaction depends upon the pollen containing the same S 'allele' as the stigma/style. It had previously been proposed that the basis of this interaction is the production of the same molecule by both pollen and stigma or style cells, and that these molecules can dimerize in an incompatible reaction to form the active molecule. However, despite considerable efforts it has not been possible to find S-RNases or SLSG in pollen. The models shown in this chapter all involve a receptor–ligand mechanism, but in each model the nature of the male determinant for self-incompatibility is unknown, and this component of the models is therefore hypothetical.

Another aspect of self-incompatibility that is also not understood is the mechanism by which different S 'alleles' are distinguished. A number of *Brassica* SLSG alleles have been sequenced and there is considerable variation between individual alleles. In addition the sequence can be divided into highly variable and less variable regions. However the important variables in determining S-locus specificity are not known.

12.7 Evolution of self-incompatibility

It is perhaps surprising that in each of the systems studied the biochemical basis of incompatibility is different. This suggests that it evolved independently in different plant groups. It may be significant that the production of self-incompatibility S-RNases in the style transmitting tissue of *Nicotiana* is accompanied by the production of a number of enzymes that would protect the style from invasion by pathogens. Further, the SRK formed in *Brassica* belongs to a family of receptor kinases, one of which acts as a disease-resistance gene in tomato. The parallel between pathogen invasion and pollen tube growth within the style, together with these findings, suggest the evolution of self-incompatibility from mechanisms of protection against pathogen attack in early angiosperms (higher plants).

Following the early evolutionary development of a self-incompatibility mechanism, the remarkable number of different S 'alleles' known to exist in self-incompatible species will arise due to the outstanding advantage any pollen containing a new S 'allele' would have within a population of self-incompatible plants.

Major Learning Objectives for Chapter 12

1. Understand the genetic basis of gametophytic and sporophytic incompatibility systems.
2. Describe the evidence that indicates that the S-locus-specific glycoproteins (SLSG) are involved in the incompatibility reaction of S 'alleles' in *Brassica*.
3. Describe the evidence that a stylar ribonuclease has a role in the incompatibility system of Solanaceae (*Petunia* and *Nicotiana*).
4. Explain how the molecular and biochemical evidence led to models for pollen and stigma or style incompatible interactions in *Brassica*, *Nicotiana* or *Petunia* and *Papaver*.
5. Be aware of the unsolved problems in understanding self-incompatibility.

Further reading

CHARLESWORTH, D. (1995). Multi-allelic self-incompatibility polymorphisms in plants. *Bioessays*, **17**, 31–38.
This review does not have an explanation of incompatibility systems but in addition to describing the molecular studies on *Brassica* and *Nicotiana*, it covers the evolution of incompatibility systems in angiosperms.

DICKSON, H. (1994). Simply a social disease? *Nature*, **367**, 517–518.
A short review that includes information about *Papaver*.

DZELZKALNS, V.A., NASRALLAH, J.B. and NASRALLAH, M.E. (1992). Cell–cell communication in plants: self-incompatibility in flower development. *Developmental Biology*, **153**, 70–82.
A full review which, with the exception of the *Papaver* example, covers the material of this chapter.

THOMPSON, R.D. and KIRCH, H-H., (1992). The S locus of flowering: when self-rejection is self-interest. *Trends in Genetics*, **8**: 381–387.
A concise illustrated review which explains gametophytic and sporophytic incompatibility systems as well as the molecular studies in *Brassica* and *Nicotiana*.

Chapter 13

Reaction of plants to changes in the environment

13.1 Signal perception and transduction: an overview

Our understanding of the response of plant cells to external stimuli is very poor and fragmentary compared to our knowledge of animal and microbial cells. Nevertheless, this is a field of critical importance to plant growth and development. Plants are static organisms and in order to develop and reproduce successfully they must be able to adjust their metabolism and growth to the constantly changing environment that surrounds them.

In this chapter, three topics will be covered: the ethylene response pathway, the abscisic acid regulatory network and the heat shock response. These have been chosen because they represent different signal perception and transduction systems that have been studied at the molecular genetic level. This group is not a complete list of topics and in fact several signal transduction pathways have already been covered in earlier chapters (see sections 6.3, 8.5, 8.6, 12.3–12.5 and Chapter 10).

Figure 13.1 shows a generalized scheme for signal detection, transduction and regulation that can be applied to signalling systems in all organisms. In this figure the external stimulus is shown as a molecule, the ligand. This binds to a receptor domain on the surface of the cell, which typically causes dimerization and conformational changes to the detector. This produces a signal that is transmitted across the cell membrane to an internal signal domain, which in turn transmits the message to intracellular components. The transduction of this signal to the response regulator usually involves an amplification stage and may also have a mechanism for integration, which will receive signals from other external stimuli.

In the transduction of a signal from an external stimulus to a cell response, it is important that the signal is removed once the external stimulus has gone. In Figure 13.1 the response regulator stimulates a termination system that switches off the signal transduction pathway. It is often important that cells react to changes in external conditions and the scheme in Figure 13.1 includes

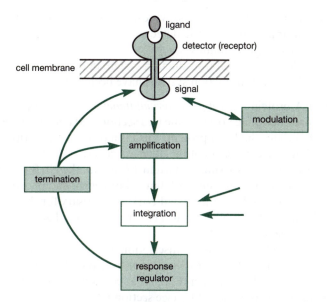

Figure 13.1 A generalized model for signal detection, transduction and regulation. In this model the signal is a molecule, the ligand, which binds to a specific dimeric receptor in the membrane causing a conformational change that produces an intracellular signal. Signal transduction to the response regulator may involve an amplification and an integration step.

a modulating mechanism which can alter the ground state activity of the signal transduction system.

Studies of signal transduction in microorganisms and animal cells have shown that there are a number of recurring themes. In order to understand the significance of the findings of preliminary studies that have been made on plant signal transduction, a short account of two major types of signal transduction mechanisms in other organisms will be given.

Prokaryote two-component signalling systems

In bacteria, signals from external stimuli are processed via protein phosphorylation. The simplest system consists of an autophosphorylating kinase (the detector–signal component) and a protein substrate of this kinase (the response regulator).

A family of autophosphorylating kinases with sequence homology have been identified in bacteria. These proteins form high-energy phosphoamino acids. The kinase autophosphorylates a conserved histidine residue, using ATP as the phosphate donor, in response to an external stimulus. This phosphoryl group is subsequently passed on to an aspartate in the response regulator. This changes the activity of this protein, which may for example be a transcriptional activator and acquire enhanced DNA binding following phosphorylation. The termination process typically involves changes in phosphatase activity, which will dephosphorylate the proteins. Modulation of the system arises from methylation of amino acids in the detector–signal, which alter this protein's sensitivity to the external stimulus.

The best-understood signal transduction system in bacteria is chemotaxis but three bacterial signal transduction systems have been introduced in this text.

1. The VirA protein from *Agrobacterium tumefaciens* is autophosphorylated in the presence of acetosyringone (see section 6.3) and subsequently phosphorylates the VirG protein. VirG is a positive transcription factor controlling the expression of the other *vir* genes.
2. In the free-living N_2-fixing bacterium *Klebsiella*, the NtrB protein acts as a kinase, in an environment with limited nitrogen, and phosphorylates the NtrC protein. Phosphorylated NtrC is necessary for the transcription of *nif*L and *nif*A and NifA in turn is a transcriptional activator of other *nif* genes (see section 8.5).
3. In *Rhizobium* NifA is also a transcriptional activator of other *nif* genes. The activity of NifA in *Rhizobium* is controlled by FixJ, which is phosphorylated by a protein (FixL) that contains a haem group and responds to the presence of O_2 (see section 8.6).

Eukaryote (animal) signal transduction via heterotrimeric G proteins

In order to illustrate the recurring themes in eukaryote signal transduction systems, two pathways involving membrane-associated trimeric G proteins will be outlined. The three subunits of the G-protein complex are known as α, β and γ. The G_α protein binds to GDP or GTP depending on its conformation; it also has GTPase activity and binds to the $\beta\gamma$ subunits in a cycle of interactions with receptor proteins and membrane-associated enzymes.

Multiple G proteins have been identified in eukaryotic organisms and each subunit can be classified according to functional or structural relationships. One of the best-studied G proteins is the G_s protein, which mediates the hormonal stimulation of adenylate cyclase. In the ground state with no extracellular signalling molecule the stimulatory G_s protein is a $\alpha\beta\gamma$ trimer associated with the cell membrane and bound to GDP (Figure 13.2, A). When the extracellular signalling molecule or ligand is present, it binds to the membrane-located receptor protein (R) and alters the conformation of this protein so that it can bind to G_s (Figure 13.2, B). Association of G_s with the ligand–receptor complex leads to a change in $G_{\alpha s}$ so that it has a weaker affinity for GDP, which is exchanged for GTP. The binding of G_s to GTP causes the dissociation of G_s into the α subunit (with GTP) and a $\beta\gamma$ dimeric subunit (Figure 13.2, C), which results in the dissociation of G_s from the ligand–receptor complex. The GTP-bound subunit now has affinity for the membrane-located enzyme adenylate cyclase, which is activated by the GTP-bound $G_{\alpha s}$ subunit. Adenylate cyclase produces cyclic AMP (cAMP) from ATP (Figure 13.2, D) and cAMP functions as a second messenger in the cell by activating the enzyme protein kinase A. Hydrolysis of GTP to GDP returns the $G_{\alpha s}$ subunit to its ground state so that it dissociates from

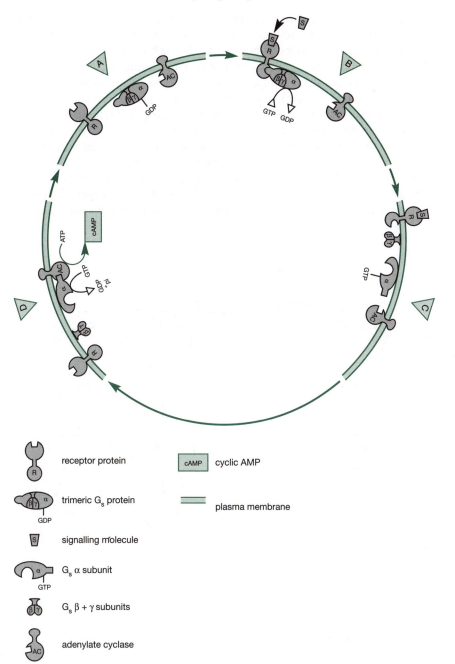

Figure 13.2 Model for the coupling of signal detection to the production of a second messenger (cyclic AMP, cAMP) through a membrane-associated heterotrimeric G_s protein. A, ground state; B, C and D, stages in the activation of adenylate cyclase by $G_{\alpha s}$ following the binding of a signalling molecule (ligand) to the receptor.

adenylate cyclase and rebinds to the βγ complex (Figure 13.2, A). The participation of G_s proteins in protein–protein interactions is thought to be brought about by diffusion of proteins within the membrane bilayer.

This signal transduction system contains two amplification steps: the production of cAMP (since many molecules of cAMP can be produced by a single activated adenylate cyclase) and the activation of protein kinase A. Protein kinase A is a serine/threonine kinase and may phosphorylate many proteins thereby activating a spectrum of other enzymes. Phosphatase enzymes are often involved in the termination of G protein-mediated signal responses by dephosphorylation of activated proteins. Integration is possible at a number of points, for example G_s molecules may bind to a range of different receptor molecules.

The activation of adenylate cyclase by G_s is also shown in Figure 13.3 where a further G protein-activated animal signal transduction pathway is illustrated.

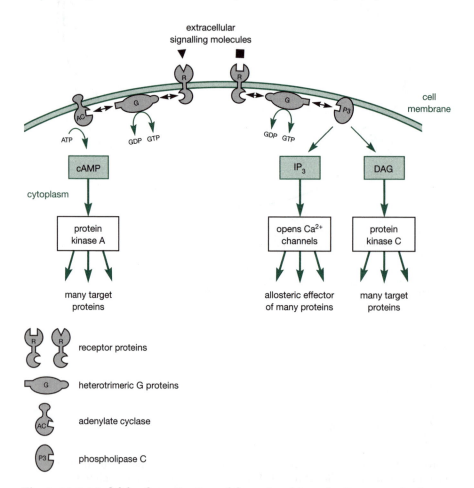

Figure 13.3 Model for the activation of three signal transduction cascades by membrane-associated heterotrimeric G proteins via the second messengers cyclic AMP (cAMP), inositol 1,4,5-trisphosphate (IP$_3$) and diacylglycerol (DAG).

The same cycle of G-protein binding to receptor, dissociation and then enzyme activation by the GTP-bound G_α subunit may also control the membrane-located enzyme phospholipase C. Phospholipase C cleaves phosphatidylinositol 4,5-bisphosphate to produce inositol 1,4,5-trisphosphate (IP_3) and 1, 2-diacylglycerol (DAG). Both of these molecules act as second messengers.

IP_3 diffuses through the cytosol to release Ca^{2+} from the lumen of the endoplasmic reticulum. A number of enzymes are activated by Ca^{2+} but the level of free Ca^{2+} ions in the cytosol is usually kept low (less than 0.2 μmol l^{-1}) because the cytosol contains large amounts of phosphate esters and calcium phosphate is insoluble. Ca^{2+} is therefore compartmentalized within the cell (typically in the lumen of the endoplasmic reticulum) or it is excluded from the cells by plasma membrane Ca^{2+} pumps. However, Ca^{2+} acts as an important second messenger for many extracellular signalling molecules, such as hormones.

DAG also acts as a second messenger. It remains in the membrane, where together with Ca^{2+} it activates protein kinase C. Protein kinase C in turn may phosphorylate membrane receptors to alter (modulate) their affinity for ligands or it may phosphorylate cellular enzymes to alter their activities.

These examples of both prokaryote and animal cell signal transduction pathways illustrate a number of features common to many systems:

1. membrane-located receptor proteins at the cell surface;
2. protein–protein interaction, coupled with allosteric conformational changes;
3. protein activation and deactivation by kinase (phosphorylation) and phosphatase (dephosphorylation) enzymes;
4. the formation of second messengers such as Ca^{2+}.

Box 13.1 Plant protein kinases

More than 70 plant protein kinase genes have been identified and this allows some general comparison to be made between protein phosphorylation regulatory mechanisms in plants and those in other organisms. Protein kinases may be classified by the amino acids that are phosphorylated into the following groups: (i) histidine kinases, (ii) tyrosine kinases, and (iii) serine and/or threonine kinases.

1. Histidine kinases are typically found in prokaryotes but recent evidence (e.g. *etr1*, see section 13.2) suggests that they may be important in plants.
2. Tyrosine kinases, including the receptor-linked tyrosine kinases (RTK), have not been found in plants.
3. The serine and/or threonine kinases have been grouped by homology.
   (a) Transmembrane receptor-like serine/threonine kinases (RLK) have been isolated from plants (e.g. *Brassica* self-incompatibility SRK) although the function of most is unknown. These may replace the important RTK group found in other eukaryotes.
   (b) The serine/threonine kinases activated by the second messenger cyclic nucleotides or calcium–phospholipids are not represented in

▶

(Box 13.1 continued)

plants but a novel group, structurally related to these, have been cloned (*in vivo* function unknown).

(c) The calcium/calmodulin-dependent and AMP-activated group appear to be important in plants, although calmodulin is usually not required for calcium activation.

Interestingly, although there are significant differences between plants and other eukaryotes in the kinases involved in the early parts of signal transduction pathways, all groups of kinase further down the phosphorylation cascade are represented in plants (including the mitogen-activated protein kinase (MAPK) cascade).

In plants there have been three important molecular approaches to the study of signal transduction. The first can be described as a search for the plant equivalents to components of animal signal transduction systems, and on the whole this approach has had only limited success. For example, there is little convincing evidence for plant signalling through inositol phosphates (e.g. IP_3) or cAMP and although G_α protein genes have been cloned (GPα1 from *Arabidopsis* and *TGα1* from tomato) other G-protein subunits have been elusive. The second approach represents a molecular study of a plant signalling system, specifically based upon the characteristics of the plant system. The incompatibility studies described in Chapter 12 illustrate this method. The third type of study involves the isolation of mutants that have an altered response to the signalling factor and these will be illustrated in sections 13.2 and 13.3.

13.2 The ethylene (ethene) response pathway

The structurally simple gas ethylene (ethene) is produced by plants both during developmental changes and in reaction to environmental factors (Figure 13.4a). It is involved in a wide range of physiological responses and in some cases it may have the opposite effects in different species. For example, in most dicotyledon species ethylene inhibits stem elongation; however in some aquatic dicotyledons and in rice it stimulates shoot growth.

The dramatic effect of ethylene on the growth of etiolated (dark-grown) dicotyledon seedlings was discovered in 1901 by Neljubov. This is known as the triple response because in dicotyledons it causes: (i) inhibition of epicotyl (or hypocotyl) and root elongation, (ii) radial swelling of epicotyl (or hypocotyl) and root cells, and (iii) loss of normal gravitropism and distortion of the apical hook.

Box 13.2 Epicotyl and hypocotyl

The base of the primary seedling stem is known as the epicotyl. In hypogeal germination the cotyledons rise above the ground due to growth elongation of the hypocotyl, which is the region between the cotyledons and the root. In

▶

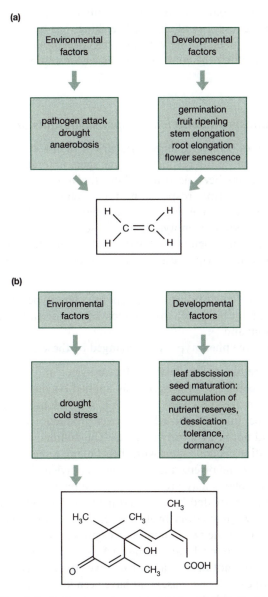

Figure 13.4 Physiological responses associated with (a) ethylene (ethene) and (b) abscisic acid.

(Box 13.2 continued)

the light the cotyledons of hypogeal germinating species differentiate into leaves (see Chapter 10).

In both types of germination the tip of the structure that pushes its way through the soil is bent into an apical hook. This probably protects the apical meristematic region or cotyledons from mechanical damage during germination.

The ethylene response pathway has been studied using two experimental approaches: the triple response of etiolated seedlings has been used as a screen to isolate mutants of *Arabidopsis* with an altered response to ethylene; the other approach is to study the control of genes whose expression is influenced by ethylene.

Ethylene biosynthesis and ethylene-response mutants in Arabidopsis

Because of the ease of screening large numbers of plants at an early stage of development and because the response is very specific to ethylene, the triple response phenotype has been used to isolate *Arabidopsis* ethylene-response mutants. Two types of triple response mutant phenotypes can be isolated: (i) seedlings that have the triple response phenotype in the absence of ethylene (constitutive triple response phenotype); and (ii) seedlings that do not have the triple response phenotype when treated with ethylene (ethylene insensitive).

Among the constitutive triple response phenotype mutants two classes can be distinguished:

1. mutants that lose constitutive triple response phenotype when grown in the presence of inhibitors of ethylene biosynthesis or antagonists of ethylene action; and
2. mutants where the phenotype is not changed by these compounds.

Class (1) mutants are plants that over-produce ethylene and therefore have the triple response in the absence of exogenously applied ethylene. The genes *eto*1, *eto*2 and *eto*3, which are shown in Table 13.1, were identified by mutations giving this phenotype. Interestingly, ethylene over-production only occurs in etiolated seedlings and not in light-grown plants. The dominant phenotype of the *eto*2 and *eto*3 mutations is consistent with their having a regulatory role in the biosynthesis of an enzyme or enzymes involved in ethylene biosynthesis. Thus *eto*2 and *eto*3 mutant plants may produce a positive regulatory protein that has lost its ability to be inactivated by a second protein. The gene *eto*1 (recessive mutation) is probably a regulatory protein also. This assumption derives from evidence showing that, of the two committed enzymes in the biosynthesis of ethylene, 1-aminocyclopropane-1-carboxylate (ACC) oxidase is constitutively expressed and *eto*1 does not map at any of the five *Arabidopsis* structural genes for the second enzyme (ACC synthase) that have been identified.

A mutation of *ctr*1 (Table 13.1) also gives a constitutive triple response phenotype but this phenotype is not affected by ethylene biosynthesis inhibitors, suggesting that this gene is involved in ethylene signal transduction. All the mutant alleles of *ctr*1 that have been isolated are recessive and pleiotropic (that is, they affect a number of characters, particularly in flower development). The recessive nature of the mutations implies loss of function and this suggests that functional *ctr*1 is a negative regulator of the ethylene response. The *ctr*1 gene has been cloned by tagging the gene with an insertion mutation caused by T-DNA. The derived amino acid sequence of the gene has the con-

Table 13.1 Ethylene biosynthesis and response mutants from *Arabidopsis thaliana*. This table does not include all of the triple response genes identified in *Arabidopsis*; those genes not included have not yet been extensively studied

Gene mutation	Inheritance	Chromosome location	Phenotype
Ethylene biosynthesis			
*eto*1	Recessive	3	Constitutive triple response phenotype; constitutive over-production of ethylene
*eto*2	Dominant	5	Constitutive triple response phenotype; constitutive over-production of ethylene
*eto*3	Dominant	3	Constitutive triple response phenotype; constitutive over-production of ethylene
Ethylene response			
*ein*1 (*etr*1)	Dominant	1	Ethylene insensitive; cloned, putative two-component kinase; reduced ethylene binding
*ctr*1	Recessive	5	Constitutive triple response phenotype; cloned, putative serine/threonine kinase
*ein*3	Recessive	3	Ethylene insensitive
*hls*1	Recessive	1	Ethylene-insensitive apical hook (hookless)

served hallmark features of mammalian serine/threonine protein kinases and this is consistent with the type of protein known to be involved in eukaryote signal transduction.

Table 13.1 includes three genes isolated by a triple response mutant screen because they are not responsive to ethylene and do not show the triple response phenotype in the presence of ethylene. The mutations *ein*1 and *etr*1 are allelic, pleiotropic and dominant. Mutant *etr*1 plants bind less ethylene than wild-type plants and this suggests that the gene may encode an ethylene receptor. The *etr*1 gene has been cloned and, as with *ctr*1, the deduced amino acid sequence has homology with proteins known to be involved in signal transduction. However, *etr*1 has clear similarity to the bacterial two-component histidine kinases. It also has three potential membrane-spanning domains. All of the *etr*1 mutant alleles isolated are dominant, as with *eto*2 and *eto*3 this mode of inheritance may indicate that protein–protein interaction is involved in *etr*1 function.

Two further ethylene-insensitive mutants are shown in Table 13.1 (*ein*3 and *hls*1) but these mutations are recessive. The mutation *ein*3 produces plants lacking all three components of the triple response phenotype but *hls*1 only affects the apical hook.

Genetic analysis of the ethylene signal transduction pathway in *Arabidopsis* using epistasis

Epistasis is a phenomenon where the mutant allele of one gene obliterates the phenotypes of the alternative alleles of another gene. This gene interaction is investigated by producing double mutants and observing the resulting phenotype. Double-mutant *ein1*(*etr1*)–*ctr1* plants have the constitutive triple response phenotype of *ctr1* plants, whereas double-mutant *ctr1*–*ein3* plants have the ethylene-insensitive phenotype of *ein3*. This epistatic relationship suggests that *ctr1* acts downstream of *ein1* and upstream of *ein3*. Figure 13.5 shows a model of the ethylene signal transduction pathway based on this analysis.

Regulation of gene expression by ethylene

Ethylene biosynthesis is stimulated by many stresses including pathogen attack and wounding, although the exact role of ethylene in plant defence against pathogens is not known. However, ethylene can elevate the steady-state mRNA levels of pathogen-related proteins (see Chapter 14), including basic chitinase and β-1,3-glucanase. Further, the application of exogenous ethylene to *etr1* and *ein3* mutant *Arabidopsis* fails to induce these genes, whereas in *ctr1* *Arabidopsis* mutants these genes are constitutively expressed.

The promoter DNA sequence element that can confer ethylene induction to a reporter gene having a minimal promoter has been identified in a number of ethylene-responsive genes. This ethylene-responsive element (ERE), GCCGCC, occurs one or more times in the bean basic chitinase gene and the tobacco β-1,3-glucanase gene. Further, four ERE-binding proteins (EREBPs), which interact directly with the GCCGCC element, have been isolated from

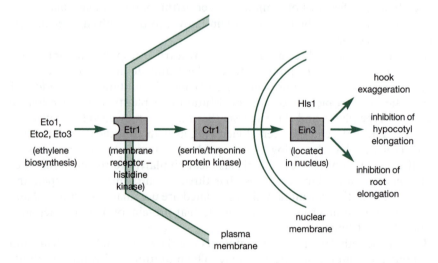

Figure 13.5 Model of the ethylene signal transduction pathway based on an analysis of *Arabidopsis* triple response mutants.

tobacco. The steady-state mRNA levels of EREBPs are themselves up-regulated by ethylene.

The EREBPs are structurally different but they all have a conserved domain of 59 amino acids, which is predicted to have DNA-binding activity. Interestingly, this domain was found to share sequence homology with domains in a number of previously identified plant genes, notably the *Arabidopsis* floral homeotic protein encoded by *Apetala2* (see Table 11.1). Genetic studies suggest that the *Apetala2* protein (AP2) may be a negative regulator of a second floral homeotic gene, *Agamous*, and although neither DNA binding nor nuclear location has been reported AP2 may directly regulate transcription of *Agamous*.

Box 13.3 Control of fruit ripening by ethylene

The other system in which the effects of ethylene have been studied at the molecular genetic level is tomato fruit ripening. Tomato has a climacteric type of fruit ripening that is controlled by ethylene and a number of mutants, which have delayed ripening, have been isolated: *Never-ripe* (*Nr*), *ripening inhibitor* (*rin*) and *non-ripening* (*nor*). The semi-dominant *Nr* is pleiotropic and does not have the ethylene-induced triple response phenotype. *Etr1* homologous genes have also been isolated from tomato and one of these is very tightly linked to *Nr*.

13.3 The abscisic acid regulatory network

Abscisic acid (ABA) is a plant hormone involved in a number of physiological responses to both environmental and developmental factors (Figure 13.4b)

During vegetative growth ABA has been proposed as an important hormone involved in triggering a wide range of physiological responses. This proposal comes partly from the observation that ABA levels increase following stress treatments such as drought, high salt or cold and partly from the fact that exogenous applications of ABA can mimic the plant responses to these environmental stimuli. The role of ABA in the plant response to drought stress and in the maturation of seeds has been extensively studied at the molecular and genetic levels. A number of mutations have been isolated that affect either the biosynthesis of ABA or the plant response to ABA. The mutations in *Arabidopsis* and maize shown in Table 13.2 commonly lead to either a wilty or viviparous phenotype. The wilty phenotype is related in part to the role of ABA in controlling stomatal opening and hence water loss from the plant via transpiration. The viviparous phenotype describes the precocious germination of seeds on the maize cob and this is related to the role of ABA in the development of seed dormancy.

In this section two approaches to the study of the ABA regulatory network will be considered: the use of ABA mutants; and studies of the control of gene expression by ABA.

Table 13.2 Abscisic acid biosynthesis and response mutants from *Arabidopsis thaliana* and *Zea mays* (maize)

Species	Gene mutation	Inheritance	Phenotype
Abscisic acid biosynthesis			
Arabidopsis	*aba*	Recessive	Epoxidation of xanthophylls
Maize	*vp2*	Recessive	Carotenoid biosynthesis
	vp5	Recessive	Carotenoid biosynthesis
	vp7	Recessive	Carotenoid biosynthesis
	vp9	Recessive	Carotenoid biosynthesis
Abscisic acid response			
Arabidopsis	*abi*1	Semi-dominant	Vegetative response to abscisic acid, stress, drought, germination; cloned, Ca^{2+}-modulated protein phosphatase
	*abi*2	Semi-dominant	Vegetative response to abscisic acid, stress, germination
	*abi*3	Recessive	Seed-specific response to abscisic acid; cloned, transcription activator, homologue of *vp*1
	*abi*4	Recessive	Seed-specific response to abscisic acid
	*abi*5	Recessive	Seed-specific response to abscisic acid
Maize	*vp*1	Recessive	Seed-specific response to abscisic acid; cloned, transcription activator, homologue of *abi*3

ABA biosynthesis and response mutants in Arabidopsis

The *Arabidopsis* mutation *aba*1 (Table 13.2) leads to a range of abnormalities that can be restored to wild type by the application of ABA. In higher plants ABA is synthesized from xanthophylls; *Arabidopsis aba* mutants are impaired in the epoxidation reaction converting zeaxanthin to antheraxanthin. The maize ABA biosynthesis mutants *vp2*, *vp5*, *vp7* and *vp9* are all blocked in earlier steps leading to the production of zeaxanthin.

Unlike these ABA biosynthesis mutations, plants with mutations in ABA response genes cannot gain the wild-type phenotype with the application of exogenous ABA. *Arabidopsis* seeds can be inhibited from germination by the application of ABA and this property has been used to select mutations leading to ABA insensitivity (Table 13.2).

The recessive mutations *abi*3, *abi*4 and *abi*5 lead to seed-specific defects whereas the semi-dominant mutations *abi*1 and *abi*2 also affect vegetative responses to ABA. The semi-dominant inheritance of *abi*1 and *abi*2 suggests a regulatory function.

Although physiological and biochemical studies demonstrate that ABA contributes to stress responses in plants, studies of ABA-responsive genes in ABA-insensitive mutant *Arabidopsis* plants indicates that the role of ABA is not straightforward and that ABA-independent pathways controlling the stress response of genes must also exist. This is illustrated in Table 13.3, which shows the expression of five *Arabidopsis* genes in wild type, *aba*1 and *abi*1 mutant plants following low temperature, drought or exogenous ABA treatment. The low-temperature treatment will acclimate the *Arabidopsis* Landsberg ecotype for frost tolerance and this physiological response is indicated in Table 13.3 as the LT_{50}, that is the temperature that kills 50% of the plants. The LT_{50} of wild-type untreated plants is –3.0°C. ABA treatment will up-regulate the steady-state mRNA levels of all these genes in wild-type plants and *aba*1 mutants, although to varying degrees. ABA, as expected, does not significantly affect gene expression in the *abi*1 mutant. However, although frost acclimation is impaired low temperature up-regulates the genes in both the *aba*1 and *abi*1 mutants. In contrast, the drought response of these genes is severely affected in both mutants.

Table 13.3 Response of *Arabidopsis* low-temperature-responsive genes in abscisic acid-null (*aba*1) and abscisic acid-insensitive (*abi*1) mutants

Gene	Low temperature			Drought			Abscisic acid		
	Wild type	*aba*1	*abi*1	Wild type	*aba*1	*abi*1	Wild type	*aba*1	*abi*1
*lti*140	+++	+++	+++	+++	NK	++	+++	++	0
*lti*30	+++	+++	+++	++	+	++	++	+++	0
*lti*45	++	++	+++	+	0	0	+	++	0
*lti*78	+++	+++	+++	+	0	+	+	+	+
adh	+++	++	++	+	0	0	++	+	0
LT_{50} (–3.0°C)	–7.4°C	–3.8°C	–5.0°C	–7.0°C	–3.0°C	–3.5°C	–6.5°C	–7.5°C	–3.0°C

Levels of expression: +++, high; ++, moderate; +, Low; 0, none; NK, not known.

The *Arabidopsis* gene *abi*1 has been cloned by identifying a cloned genomic sequence containing flanking RFLP markers and then producing transgenic plants containing candidate genes. Because the *abi*1 mutation is semi-dominant, the mutation can be recognized by a phenotype in transgenic plants. The wild-type *abi*1 protein (ABI1) is predicted to have two domains. The N-terminal domain has the structural elements of an 'EF hand' Ca^{2+}-binding protein and the C-terminal domain has homology with a class of serine/threonine protein phosphatase enzymes. This suggests that ABI1 may be a novel Ca^{2+}-dependent phosphatase. The involvement of Ca^{2+} and phosphatase/kinase enzymes in an ABA signal transduction pathway is not unexpected from our knowledge of signal transduction in animals (see section 13.1). In addition, changes in cystolic Ca^{2+} have been observed in a range of plant cells following phytohormone (including ABA) and environmental stimuli (Table 13.4).

Table 13.4 Stimuli inducing changes in cytosol Ca^{2+} levels in higher plants

Treatment	Plant	Organ/tissue	Change in Ca^{2+} level
Abscisic acid	Various	Guard cells	Transient increase
(phytohormone)	Parsley	Hypocotyl	Transient increase
	Parsley	Root	Transient increase
	Maize	Coleoptile	Transient increase
	Maize	Root	Transient increase
Auxin (phytohormone)	Maize	Epidermis	Oscillations
	Maize	Coleoptile	Sustained increase
	Maize	Roots	Sustained increase
	Parsley	Hypocotyl	Sustained increase
	Parsley	Roots	Sustained increase
Gibberellic acid (phytohormone)	Various	Aleurone cells	Sustained increase
Cold shock	Tobacco	Seedling	Transient increase
Wind	Tobacco	Seedling	Transient increase
Touch	Tobacco	Seedling	Transient increase
Pathogen elicitors	Tobacco	Seedling	Transient increase

The *abi*3 gene from *Arabidopsis* has also been cloned by chromosome walking. The deduced amino acid sequence of the *abi*3-encoded protein (ABI3) has homology with the maize VP1 protein, which functions as a transcription activator. This is confirmed by over-expression of ABI3 in transgenic plants, which will *trans*-activate an ABA-inducible reporter gene.

ABA signal transduction pathway

The production of double mutants in *Arabidopsis* has also been used to study the epistatic interactions of mutant alleles from different *abi* genes. This analysis shows that *abi*3 or *abi*5 combined with either *abi*1 or *abi*2 has additive effects. However, the *abi*1–*abi*2 and the *abi*3–*abi*5 double mutants are epistatic and the resulting phenotype resembles that of a single *abi* mutation. The *abi*4 mutation is distinct because it is epistatic with both *abi*2 and *abi*3 and this suggests a network of signal transduction rather than single ABA signal pathways involving *abi*1 and *abi*2 in vegetative tissue and *abi*3 and *abi*5 in seeds. A model for ABA signal transduction incorporating the known facts is shown in Figure 13.6. Although an ABA-receptor protein has not been isolated, ABA has been shown to bind to the outside of the plasma membrane and signal intracellular changes from outside the cell; therefore a membrane-located ABA receptor is included in Figure 13.6.

Regulation of gene expression by ABA

The ABA signal transduction network is also being investigated by studying the expression of ABA-inducible genes. Analysis of the promoters of these genes

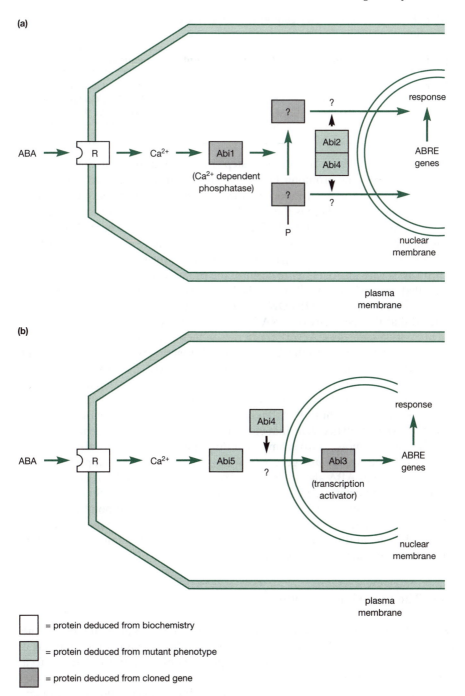

Figure 13.6 Model of the abscisic acid (ABA) signal transduction pathways in (a) vegetative and (b) seed tissues based on an analysis of *Arabidopsis* abscisic acid-insensitive mutants. R, putative ABA receptor protein; ABRE genes, genes responsive to ABA; P, phosphate.

Table 13.5 Promoter DNA sequence elements controlling the abscisic acid induction of abscisic acid-responsive genes

Species	Gene	Promoter element	Sequence
Maize	*Em*	Em1A	GGACACGTGGC
Rice	*rab*16A	Motif 1	CCGTACGTGGCGC
Arabidopsis	*RD*29A	DRE	TACCGACAT
Maize	*C*1	Sph	TCCATGCATGCAC

gives information about the terminal steps in the signal transduction network. Table 13.5 shows the promoter DNA sequence elements that have been shown to be involved in the ABA-induced expression of ABA-responsive genes. The best characterized of these is the Em1A element from maize, which shares the sequence ACGTGGC in motif 1 of the rab16A rice gene. The core of these two elements (which have been named abscisic acid-responsive elements, ABRE; see Table 1.2) is the C/T ACGTG palindrome known as the G-Box.

Sequences with an ACGT core occur in the promoters of a wide range of plant genes that are environmentally controlled and this core is involved in the binding of basic leucine zipper (bZIP) DNA-binding protein (transcription factor). It is believed that the specificity of DNA binding by these proteins depends on flanking nucleotide sequences. A bZIP transcription factor (EmBP-1) has been isolated from maize which binds to the *Em* gene ABRE together with a calcium-binding protein (GF-14), which also seems to be part of the ABRE DNA–protein complex. The protein GF-14 has homology to the 14-3-3 class of mammalian kinases. Although experimental evidence from *in vivo* promoter–reporter gene constructs, as well as *in vitro* gel shift DNA-binding assays, supports the role of the G-Box ABRE in ABA-mediated gene induction, not all ABA-responsive genes have this motif in their promoter. Two examples of this type of promoter are also shown in Table 13.5. The *Arabidopsis* gene *RD*29A is responsive to low temperature and drought as well as ABA and its drought-responsive element (DRE) was identified by promoter–reporter gene constructs in transgenic plants. The Sph element in maize *C*1 promoter was identified by analysis of deletion and point mutations. The results of these promoter studies also demonstrate the probable complexity of the ABA signal transduction network, which may involve a number of different transcription-activating mechanisms.

13.4 The heat-shock response

One of the best-characterized plant responses to an environmental factor is the heat-shock response. This is a response to an environmental stress that has a role in survival, since the heat-shock response confers thermotolerance to a subsequent normally lethal temperature.

The heat-shock response is observed in almost every living organism, including prokaryotes. Because of the high level of conservation of many of the proteins involved, information from other organisms can be cautiously applied to the plant response. The heat-shock response has a similar pattern in all organisms:

1. cessation of normal protein and mRNA synthesis;
2. *de novo* synthesis of a conserved set of heat-shock proteins (HSPs);
3. preservation of existing mRNA;
4. acquisition of thermotolerance;
5. gradual decline in HSP synthesis with prolonged heat treatment;
6. HSP induction by arsenite.

The HSPs (Figure 13.7) are encoded by members of multigene super-families and, with the exception of the low molecular weight group (LMW HSPs), there are members of these multigene families that are constitutively expressed and not heat-shock responsive. In many cases, it has been the study of these heat-shock protein cognates (HSCs) that has provided information about protein function. Table 13.6

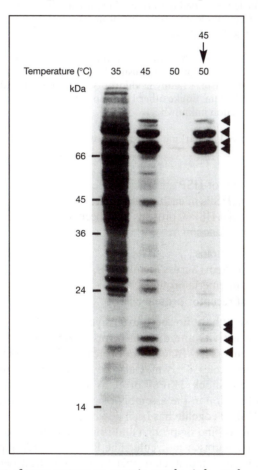

Figure 13.7 Effect of temperature on protein synthesis by 40-hour-old seedlings of the tropical grain, pearl millet (*Pennisetum typhoideum* Rich.). Seedlings were incubated in ^{35}S-methionine for 2 hours at 35°, 45° or 50°C or at 50°C following a 30-min pretreatment at 45°C. Proteins were separated by sodium dodecyl sulphate polyacrylamide gel electrophoresis and visualized by fluorography. Arrows indicate heat-shock proteins. (From Ougham, H.J. and Howarth, C.J. (1988) Temperature shock proteins in plants. In Long, S.P. and Woodward, F.J. (eds) *Plants and temperature*, pp. 259–280. Company of Biologists, Cambridge.)

Table 13.6 Classification and properties of heat-shock proteins (HSPs)

HSP family	Properties
HSP100	Not extensively studied in plants; HSP104 important for thermotolerance in yeast
HSP90	No specific studies in plants
HSP70	Most conserved group of HSP; have ATPase activity; distinct forms found in the cytoplasm, endoplasmic reticulum lumen, mitochondria and chloroplasts; involved in protein folding and transport
HSP60	Nucleus encoded but present in mitochondria and chloroplasts (includes Rubisco-binding protein); involved in assembly of oligomeric proteins
Low molecular weight HSP	Molecular weight varies between 17 and $28 \times 10^3 M_r$, more prominent in plants than other organisms; four multigene families: two families in cytoplasm, 1 family in chloroplasts, 1 family in endoplasmic reticulum; unlike other HSPs no constitutively expressed members; function unknown
Ubiquitin	Involved in marking denatured proteins for proteolytic degradation

lists the different types of HSP and summarizes the properties of these proteins. The exact size of the HSPs in each group varies a little between different species. The HSP90, HSP70 and HSP60 proteins have been termed 'molecular chaperones'. They are proteins that function in:

1. folding other proteins,
2. transport of proteins across membranes,
3. assembly of oligomeric proteins,
4. modulation of receptor protein activity.

HSP100 and HSP90 have not been extensively studied in plants. HSP70 proteins have ATPase activity and distinct forms of HSP70 are found in different cellular compartments. They function in protein folding and in the transport of proteins across membranes. They may also be involved in the regulation of HSP transcription by heat shock. HSP60 proteins are present in mitochondria and chloroplasts and function in the assembly of oligomeric proteins. One constitutively expressed HSP60 cognate has been described in section 3.5. This protein is known as the Rubisco-binding protein and is involved in the correct assembly of the L_8S_8 multimeric Rubisco molecule. The Rubisco-binding protein, which is nuclear encoded, is an abundant chloroplast protein in the absence of heat stress and there is no evidence that levels of the protein increase following heat stress.

There are four multigene families encoding LMW HSPs in plants and in contrast with other organisms they are often the most abundant HSPs. There is evidence that each LMW HSP family is compartmentalized, with two found in the cytoplasm and the others occurring in the mitochondria and the chloroplasts. There are no LMW HSCs and the function of the LMW HSPs is not known.

Typically a temperature about 5°C higher than the optimum for growth will elicit the heat-shock response and the synthesis of HSP. However, it is not possible to describe exactly how any HSP contributes to thermotolerance. Most of the evidence for an HSP function in thermotolerance is correlative and the only HSP for which there is good genetic evidence is HSP104 in yeast. In plants, genetic differences in thermotolerance exist but studies of its inheritance have not shown that this is associated with differences in the HSP profile. These genetic studies are complicated by the complexity and number of HSP genes and by the fact that the parents used often differed at a number of other genes (that is the genetic background of the HSP variation was different).

Regulation of HSP gene expression

The mechanism of temperature perception is not known in plants but in other organisms experimental and genetic evidence suggests that HSC70 (a constitutively expressed member of the HSP70 group) plays a central role.

Using reporter genes in transgenic plants the important sequence motifs of the HSP promoters have been extensively studied in plants. The *cis*-acting heat-shock element (HSE) consists of the consensus palindromic sequence -GAA-TTC- but the minimal effective sequence must have 1.5 HSEs and many HSP genes have seven. In addition distal A/T-rich regions, SAR and a CCAAT box have been shown to affect the level of HSP gene transcription. Genes encoding proteins that bind to the HSE have been isolated from plant expression libraries using the consensus HSE as a probe. These are called heat-shock factors (HSF). In other organisms HSF binding to HSEs is stimulated by trimerization of the HSFs, which in turn is inhibited by HSP70.

These findings lead to the model of activation of HSP gene transcription shown in Figure 13.8. The transcription activators (HSF) are shown bound to a constitutive HSP70 (HSC70) in the absence of heat. Heat stress causes the dissociation of HSP70 from HSFs, which are then able to oligomerize to form trimers and bind to the HSE in the HSP promoter. This activates RNA polymerase II and the HSP gene is transcribed. There is some evidence that phosphorylation of HSF increases the level of transcription. Other proteins are thought to be involved in the activation of HSP transcription and these can be expected to interact with other promoter sequence features such as the A/T-rich region and the CCAAT box.

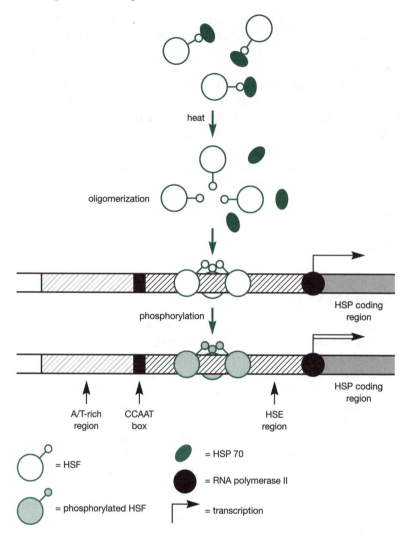

Figure 13.8 Model showing the interaction of the constitutive $70 \times 10^3 \; M_r$ heat-shock protein (HSP70), the heat-shock transcription factor (HSF) and the heat-shock promoter elements (HSE) during the heat-shock induction of heat-shock proteins (HSP).

Major Learning Objectives for Chapter 13

1. Recognize and identify common features of signal transduction pathways that have been found in plants.
2. Understand how epistasis provides information about the relative position of gene action in a signal transduction pathway.
3. Explain how the genetic, molecular and biochemical evidence led to models for ethylene and abscisic acid signal transduction in plants.

(Major Learning Objectives for Chapter 13 continued)

4. Describe the evidence that distinguishes mutations affecting either ethylene or abscisic acid biosynthesis from mutations in the signal transduction pathways.
5. Understand the evidence indicating that abscisic acid signal transduction involves a network of activating mechanisms.
6. Be acquainted with the conservation of heat-shock response, which allows data from other organisms to be applied to plant heat shock proteins.

Further reading

The study of ethylene signal transduction using *Arabidopsis* mutants is covered by the following two references.

ECKER, J.R. (1995). The ethylene transduction pathway in plants. *Science*, **268**, 667–675.

KIEBER, J.J. and ECKER, J.R. (1993). Ethylene gas: it's not just for ripening any more! *Trends in Genetics*, **9**, 356–362.

ROCK, C.D. and QUATRANO, R.S. (1994). Insensitivity is in the genes. *Current Biology*, **4**, 1013–1015.
This short review article summarizes the analysis of abscisic acid signal transduction using *Arabidopsis* mutants.

VIERLING, E. (1991). The roles of heat shock proteins in plants. *Annual Reviews of Plant Physiology and Plant Molecular Biology*, **42**, 579–620.
A comprehensive account of plant heat-shock proteins.

Chapter 14

The molecular biology of disease and pest resistance

14.1 Plant–pathogen interactions

Plant pathogens include viruses, bacteria and fungi, although the predominant plant pathogens are fungi. However most plant species are unable to serve as host for most species of viruses, bacteria and fungi, which are therefore non-pathogenic. Where a microorganism can invade a plant to become pathogenic there must be an interaction between plant and microorganism. The interaction will parallel the kind of interaction described for plants and *Agrobacterium* (crown gall disease) in Chapter 6 and for the symbiotic relationship between legumes and *Rhizobium* spp. in Chapter 8.

Molecular studies of the interaction of plants and pathogens fall into two types. Either they represent studies based on an analysis of the response of the plant (induced genes) to pathogen attack or they are studies based on intraspecific plant variation in resistance to a particular pathogen. This chapter will focus primarily on the second approach.

Table 14.1 shows a simple classification of the relationship between variation in the plant host and variation in the pathogen which leads to the formation of disease symptoms. This simple table, which ignores partial resistance, shows that the plant may be classified as either resistant or susceptible and the pathogen as either virulent or avirulent. The only combination that produces disease symptoms is a virulent pathogen invading a susceptible plant. Plant variation in resistance to pathogens can be classified into two types:

1. variation arising from alleles of a single gene (sometimes called vertical resistance), which is usually pathogen race-specific; and
2. variation arising from the combined effects of alleles of a number of genes (multigenic), sometimes called horizontal resistance, broad resistance or field resistance, which is usually pathogen race-non-specific.

Molecular studies are largely devoted to the single-gene control of disease resistance. A small number of plant resistance genes and pathogen virulence

Table 14.1 Interaction between pathogen and potential host plant

Host (plant)	Pathogen	
	Virulent	Avirulent
Resistant	No disease	No disease
Susceptible	Disease	No disease

genes have now been cloned. Before this molecular work is described, a brief general account of the mechanisms of pathogen resistance in plants will be given. Clearly, the study of plant pathogens is a very large topic and this section presents a brief overview in order to place the molecular work in context.

Resistance of individual plant genotypes to a pathogen may arise from a number of mechanisms: (i) preformed chemical inhibitors, (ii) induced chemical inhibitors, (iii) structural barriers, and (iv) rapid necrosis with systemic acquired resistance.

Preformed chemical inhibitors

An example of the role of a preformed chemical inhibitor comes from the resistance of green tomato fruits to the fungus *Fusarium*. Green tomato fruits contain a steroidal glycoalkaloid, tomatine, and are resistant to *Fusarium*. In order to test the role of this alkaloid in resistance, mutant *Fusarium* strains were selected that could grow on high levels of tomatine. These strains were shown to have increased virulence on green fruits and this implicates tomatine in the resistance of immature fruits to the fungus.

Induced chemical inhibitors

A number of studies have shown that pathogen attack will induce a range of metabolic changes in the host plant including the synthesis of pathogen-related proteins. Most of the data implicating these changes in resistance mechanisms are correlative. However, one class of secondary plant metabolites, the phytoalexins, has been assessed more critically.

The fungal pathogen of pea, *Fusarium solani* f. sp. *pisa*, produces an enzyme (demethylase) that can detoxify the pea phytoalexin pisatin. The gene from *Fusarium* encoding this enzyme has been cloned and transformed into the fungus *Cochliobolus heterostrophus*, which is a pathogen of maize but not of peas. *C. heterostrophus* transformants were found to produce significant growth on peas compared with the wild-type non-transformant strain. This suggests that a single phytoalexin can confer significant resistance to fungal pathogens in peas. However, the lesions caused by *C. heterostrophus* are limited and the transformed fungus cannot invade the whole plant. This indicates that the pathogenicity of *Fusarium* in peas involves other genes.

Structural barriers

Structural variation between genotypes can be a major determinant of resistance to pathogens. In barley, spores of the fungal disease loose smut (*Ustilago nuda* f. sp. *hordei*) are spread by wind. They enter open flowers and infect the plant via the stigma and ovary wall. Barley is a naturally self-pollinating species in which pollination usually occurs before the flowers open. Since pollination does not require open flowers, a major goal of European barley breeders has been the development of a barley variety with closed flowers that would as a consequence be resistant to loose smut.

Rapid necrosis and systemic acquired resistance

An important active defence system in plants is called the hypersensitive response (HR), which is characterized by rapid necrosis of the cells in the vicinity of the invading pathogen. This resistance mechanism is thought to include two phases: firstly, the death of the surrounding cells may limit the growth of a pathogen; secondly, there is a broad physiological immunity, called systemic acquired resistance (SARes), that results from the HR.

It has been shown that HR causes increased levels of salicylic acid in the plant and that exogenous applications of salicylic acid induce the same set of SARes genes that are induced by biological SARes. These findings led to the idea that salicylic acid is an endogenous signal for SARes.

Recently transgenic tobacco plants have been produced that express the salicylate hydroxylase gene (*nah*G) from the bacterium *Pseudomonas putida*. This enzyme converts salicylic acid to catechol, which is inactive in SARes. The *nah*G transgenic tobacco plants do not accumulate salicylic acid following pathogen attack and have no SARes. This result implicates salicylic acid in the development of SARes.

In view of the impact that plant diseases have on crop yield, the study of disease resistance is a major topic in plant science. A large number of genes affecting disease resistance have been identified by classical genetic analysis. This is illustrated in Table 14.2, which shows the loci that confer fungal disease resistance identified in barley. These loci are distributed among all seven barley chromosomes. For many diseases, several loci (and therefore genes) give rise to resistance; thus there are 13 resistance loci, with about 92 alleles, known for powdery mildew of barley. This multiplicity of genes is consistent with the idea that a wide range of different resistance mechanisms exist in barley.

14.2 Classical studies of the genetics of plant–pathogen interaction in flax (*Linum usitatissimum*)

The pattern of the resistance of flax (*L. usitatissimum*) to flax rust was elucidated in the 1940s and 1950s by Flor, who analysed crosses between resistant

Table 14.2 Genetic loci conferring fungal disease resistance in barley (not all the loci have been located to a chromosome)

Disease	Number of loci	Chromosomes
Leaf rust (*Puccinia hordei*)	11	1, 2, 3, 5, 6
Stem rust (*P. graminis*)	2	1
Stripe rust (*P. striiformis*)	4	5
Loose smut (*Ustilago nuda*)	7	1, 5
Covered smut (*U. hordei*)	4	
Semi-loose smut (*U. nigra*)	1	
Powdery mildew (*Erysiphe graminis* f. sp. *hordei*)	13	4, 5, 6
Net blotch (*Pyrenophora teres*)	4	2, 3, 5
Barley stripe (*P. graminae*)	3	
Spot blotch (*Cochliobolus sativus*)	4	2, 5
Septonia blotch (*Leptosphaeria avenaria* f. sp. *triticea*)	3	
Scald (*Rhynchosporium secalis*)	11	3, 4
Scale (*Fusarium* spp.)	1	
Blast (*Pericularia oryzae*)	1	

and susceptible varieties of flax and crosses between virulent and avirulent races of the pathogen (rust).

Box 14.1 Flax rust (*Melampsora lini*)

Flax rust is a basidiomycete fungus. It is an obligate parasite (it cannot grow in axenic culture) and, in common with other basidiomycetes, it has a biphasic life cycle. During part of the life cycle it grows as a monocaryon, that is the hyphal cells contain a single haploid nucleus. In the second phase of the life cycle, the fungus is a dicaryon and the hyphal cells contain two haploid nuclei. Each of the pair of haploid nuclei comes from a different parent monocaryon, which may be genetically different. The presence of two haploid nuclei in each cell means that the fungus is essentially diploid. *Melampsora* genes have a diploid pattern of inheritance and the alleles of individual genes may be dominant or recessive. The paired haploid nuclei only fuse to form a diploid nucleus immediately prior to meiosis.

The inheritance of pathogen virulence and the inheritance of flax resistance to two races of rust is illustrated in Tables 14.3 and 14.4. Table 14.3 demonstrates the pattern of inheritance of virulence versus avirulence in the progeny of two races of rust (22, 24). Table 14.4 shows the pattern of inheritance of resistance versus susceptibility in the progeny of two varieties of flax (Ottawa, Bombay).

Table 14.3 shows that race 22 of the rust is virulent on Ottawa but not on Bombay, whereas race 24 is virulent on Bombay but not on Ottawa. The F_1 hybrid is avirulent on both flax varieties. Races 22 and 24 are genetically different and these results indicate that the allele for virulence in each race is recessive (since the F_1 is avirulent on both varieties). Intercrossing F_1 cultures produces an F_2 generation, also shown in Table 14.3. Inspection of the F_2 progeny shows individuals virulent or avirulent on Ottawa and virulent or

Table 14.3 Inheritance of virulence in the progeny of two races (22, 24) of flax rust (*Melampsoro lini*). (From Flor, H.H. (1956) The complementary genic systems in flax and flax rust. *Advances in Genetics*, **8**, 29–54)

Host (flax variety)	Race		F₁	F₂			
	22 ($a_La_LA_NA_N$)	24 ($A_LA_La_Na_N$)	($a_LA_La_NA_N$)	(A_L-A_N-)	($a_La_LA_N$-)	(A_L-a_Na_N)	($a_La_La_Na_N$)
Ottawa (*LLnn*)	Susceptible	Resistant	Resistant	Resistant	Susceptible	Resistant	Susceptible
Bombay (*llNN*)	Resistant	Susceptible	Resistant	Resistant	Resistant	Susceptible	Susceptible
No. of cultures				78	27	23	5

Genotypes in parentheses: L, resistant to A_L; l, susceptible to A_L; N, resistant to A_N; n, susceptible to A_N; a_L, virulent on L; A_L, avirulent on L; a_N, virulent on N; A_n, avirulent on N; dash indicates unknown allele.

Table 14.4 Inheritance of resistance to two races of flax rust (22, 24) in the progeny of two varieties (Ottawa, Bombay) of flax (*Linum usitatissimum*). (From Flor, H.H. (1956) The complementary genic systems in flax and flax rust. *Advances in Genetics*, **8**, 29–54)

Race	Variety (host)		F₁	F₂			
	Ottawa (*LLnn*)	Bombay (*llNN*)	(*LlNn*)	(*L-N-*)	(*L-nn*)	(*llN-*)	(*llnn*)
22 ($a_La_LA_NA_N$)	Susceptible	Resistant	Resistant	Resistant	Susceptible	Resistant	Susceptible
24 ($A_LA_La_Na_N$)	Resistant	Susceptible	Resistant	Resistant	Resistant	Susceptible	Susceptible
No. of plants				110	32	43	9

Genotypes in parenthesis, L, resistant to A_L, l, susceptible to A_L, N, resistant to A_N, n, susceptible to A_N, a_L, virulent on L; A_L, avirulent on L; a_N, virulent on N; A_n, avirulent on N; dash indicates unknown allele.

avirulent on Bombay. Further, the progeny can be classified into four types of individual: avirulent on both varieties, only virulent on Ottawa, only virulent on Bombay, or virulent on both varieties, with the number of cultures in each occurring in a 9:3:3:1 ratio. This Mendelian F_2 ratio is expected when the alleles of the two genes segregate independently during meiosis and indicates that the genes are not linked. This result demonstrates that (i) two different genes control virulence on the flax varieties Ottawa and Bombay and (ii) these two genes are not linked in the rust genome. On the basis of this and similar analyses, race 22 has been given the genotype, $a_L a_L A_N A_N$ and race 24 has been given the genotype $A_L A_L a_n a_n$. A_L/a_L and A_N/a_N represent different genes with the allele a_L conferring virulence on Ottawa and a_N conferring virulence on Bombay.

Table 14.4 shows the results of crossing the flax variety Ottawa with the variety Bombay to produce an F_1 and an F_2 generation. These plants have been tested for their susceptibility or resistance to the rust races 22 and 24. As seen from Table 14.3 Ottawa is susceptible to race 22 but resistant to race 24, whereas Bombay is susceptible to race 24 but resistant to race 22. The F_1 progeny produced by crossing Ottawa and Bombay are resistant to both races. Ottawa and Bombay are genetically different and, since the F_1 plants are resistant to both races, these data show that the allele for resistance in each variety is dominant. Intercrossing F_1 plants produces an F_2 generation that can be classified into four types: plants resistant to both races, plants resistant to race 24 only, plants resistant to race 22 only and plants susceptible to both races, with the numbers in each producing a 9:3:3:1 ratio. This result is also a Mendelian dihybrid F_2 segregation ratio indicating that (i) two different genes are responsible for flax resistance to race 22 and race 24, and (ii) the gene for resistance to race 22 is not linked to the gene for resistance to race 24 in the flax genome.

The flax variety Ottawa has been assigned the genotype *LLnn* and the variety Bombay the genotype *llNN*. The allele *L* confers resistance to the A_L rust genotype and the allele *N* confers resistance to the A_N rust genotype. From Tables 14.3 and 14.4 the interaction between rust and plant genotypes can be summarized as shown in Table 14.5. This pattern of single gene resistance and virulence is known as gene-for-gene interaction, and it has been described for a number of viral, bacterial and fungal pathogens that elicit an HR in resistant plants. One feature of the gene-for-gene interaction, illustrated by flax rust resistance, is that the virulence alleles are recessive. Such alleles arise from loss of function mutations; this is counter-intuitive, since it indicates that the functional fungal gene gives rise to the avirulent phenotype (this is an incompatible type of interaction described in section 14.3).

The type of resistance demonstrated by the flax example can be very effective in protecting the plant from infection and has been used widely in plant breeding programmes. However, because loss of function mutations leads to virulence and because the population size of the pathogen is much larger than the plant, this type of resistance is unstable, with new pathogen mutants arising within a few years of the introduction of the resistance allele into the crop.

Table 14.5 Summary of interactions between rust and plant genotypes

	Plant resistance gene L			Plant resistance gene N	
Rust alleles	**L**	**l**	**Rust alleles**	**N**	**n**
A_L	Resistant	Susceptible	A_N	Resistant	Susceptible
a_L	Susceptible	Susceptible	a_N	Susceptible	Susceptible

Box 14.2 Resistance genes in flax

There are a total of five loci (K, L, M, N, P) known in flax that confer resistance to a single pathogen, rust flax. Locus L, described in this section, appears to be a multi-allelic series of one gene, whereas locus M is probably a multigene family.

14.3 Molecular genetics of plant disease resistance: HR

Since 1992, a number of plant resistance genes have been cloned and these are listed in Table 14.6 together with the pathogen and pathogen virulence gene involved in the gene-for-gene interaction.

Incompatible interactions

The interaction between tomato and the leaf mould fungus *Cladosporium fulvum* is equivalent to the interaction between flax and *Melampsora lini* (flax rust) and is shown diagrammatically in Figure 14.1a. The avirulence (functional) gene (*Avr9*) of *Cladosporium* has been shown to encode a peptide that acts as signal molecule (elicitor) triggering the HR in tomato. The corresponding resistance gene in tomato, *Cf-9*, is thought to encode a leucine-rich receptor molecule (LRR) for the Avr9 peptide. The interaction of the Avr9 peptide and *Cf-9* LRR receptor probably causes a conformational change in the receptor and this may act as the first signal for the plant cell HR. With this mechanism the only combination of alleles that will elicit HR in tomato is a functional (dominant) *Cf-9* allele in tomato and a functional *Avr9* allele in *Cladosporium* (Figure 14.1a). In all the other combinations either the elicitor or the receptor or both are missing, the HR is therefore not triggered and the plant is susceptible to infection.

This receptor–elicitor (ligand) model of the classic gene-for-gene interaction has been termed an incompatible interaction. It also describes the interaction of the tomato *PTO* resistance gene and *Pseudomonas syringae avrPto* avirulence gene, the interaction of the *Arabidopsis RPS2* resistance gene and the *P. syringae avrRpt2* avirulence gene, and the interaction of the tobacco *N'* gene and tobacco mosaic virus, as well as flax resistance genes and flax rust (Table 14.6).

Table 14.6 Cloned plant genes that confer resistance to pathogens

Plant species	Plant resistance gene	Protein encoded by resistance gene	Pathogen species	Pathogen virulence/ avirulence gene	Product of virulence/ avirulence gene	Method of cloning	Interaction (genes in pathogen)
Tomato	*Cf-9*	Contains leucine-rich repeats	*Cladosporium fulvum* (ascomycete)	*Avr9*	Peptide	TT	Incompatibility genes
Maize	*HM1*	NADPH-dependent HC-toxin reductase	*Cochliobolus carbonum* (ascomycete)	*Tox2* locus	HC-toxin	TT	Compatibility genes
Tomato	*PTO*	Serine threonine kinase	*Pseudomonas syringae* pv. *tomato* (bacterium)	*avrPto*	(Syringolides[a])	MAP	Incompatibility genes
Arabidopsis	*RPS2*	Contains leucine-rich repeats	*P. syringae* pv. *tomato* and *maculicola* (bacterium)	*avrRpt2*	(Syringolides[a])	MAP	Incompatibility genes
Tobacco	*N''*	Contains leucine-rich repeats	Tobacco mosaic virus	*CP*	Coat protein	TT	Incompatibility genes
Flax	*L⁶*	Contains leucine-rich repeats	*Melampsora lini* (basidomycete)	a_L6	Not known	TT	Incompatibility genes

[a] Syringolides are the products of an avirulence gene in *P. syringae* pv. *glycinea*.

TT, transposon tagging; MAP, map-based cloning (i.e. chromosome walking from linked restriction fragment length polymorphism).

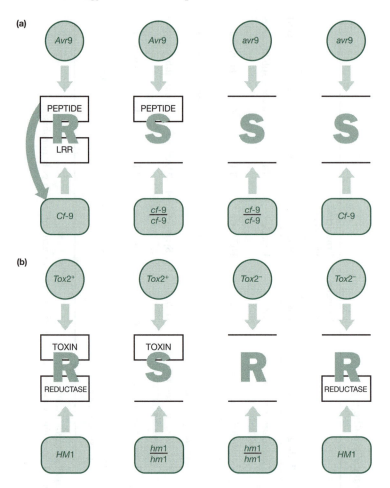

Figure 14.1 Plant–pathogen interactions that specify plant disease resistance or susceptibility. (a) Interaction of tomato (*Lycopersicon esculentum* Miller) *Cf*-9 resistance gene and the *Cladosporium fulvum Avr*9 avirulence gene, an example of a pathogen incompatibility gene. (b) Interaction of maize (*Zea mays* L.) *HM*1 resistance gene and *Cochliobolus carbonum Tox*2+ gene, an example of a pathogen compatibility gene. See text for details. R, plant resistant to pathogen; S, plant susceptible to pathogen.

Box 14.3 Function of *Cladosporium fulvum Avr*9 gene

A tomato genotype lacking a functional *Cf*-9 gene was transformed with the *Cladosporium Avr*9 functional allele under the control of a plant promoter. When the transgenic tomato was crossed with a line containing functional *Cf*-9 alleles and a transposable element (Ds), most of the progeny died because interaction of the *Avr*9 gene product with the *Cf*-9 gene product led to cell death via the HR. However, some of the plants survived and these were found to be Ds-inactivated mutants of the *Cf*-9 gene.

The question of the role of those pathogen genes whose function leads to avirulence on some plant genotypes has not been established for fungi. In bacteria, it has been shown that they may lead to enhanced virulence on susceptible hosts compared with strains containing the non-functional allele.

The plant resistance gene-encoded receptors, predicted by this receptor–ligand model, are thought to activate a signal transduction pathway leading to HR. Established cellular reactions in the HR include the production of reactive oxygen intermediates (such as O_2^-, H_2O_2, $OH^{\cdot}$), alterations in K^+ efflux, phytoalexin production and changes to the cell wall.

Compatible interactions

Figure 14.1b illustrates another type of pathogen–plant interaction. The fungus *Cochliobolus carbonum* (previous name *Helminthosporium carbonum*) causes leaf blight and ear mould of maize. Race 1 of *C. carbonum* produces a cyclic tetrapeptide toxin (called HC-toxin) and the production of this toxin co-segregates with a single locus, $Tox2^+$. The fact that application of pure HC-toxin in conjunction with a non-infectious isolate of *C. carbonum* permits infection establishes HC-toxin as a compatibility factor for this fungal pathogen. The target, within the plant, for the toxin is not known and so the mechanism by which the toxin enables infection is also not known.

Some genotypes of maize are resistant to infection by race 1 of *C. carbonum*. This resistance is controlled by a nuclear gene, *HM*1, which is located on chromosome 1. Dominant (functional) alleles of *HM*1 provide full resistance to race 1 of *C. carbonum* and this resistance is caused by the production of an enzyme (HC-toxin reductase), encoded by *HM*1, which inactivates the HC-toxin.

Figure 14.1b shows that the only interaction that leads to infection is a functional $Tox2^+$ allele in the fungus and the absence of a functional *HM*1 allele in the plant. Loss of function mutations in the pathogen lead to avirulence on all genotypes of the plant. This means that *HM*1 resistance in maize to the compatibility gene in *C. carbonum* is more stable than plant resistance to incompatibility genes in pathogens (illustrated in Figure 14.1a).

14.4 Characterization of cloned plant HR resistance genes

Analysis of the amino acid sequences, deduced from the DNA sequences of the cloned plant resistance genes shown in Table 14.6, gives some information about the proteins encoded by these genes.

Maize *HM*1 encodes an HC-toxin reductase and was discussed in the previous section. Tomato *PTO* encodes a protein with homology to a group of serine/threonine (protein) kinases. The other genes (*RPS2*, *N'*, *Cf*-9, and *L*^{6}) are all predicted to encode proteins containing leucine-rich repeats (LRR); these four proteins are structurally quite different from both HM1 and PTO. Although the RPS2, N', Cf-9 and L^6 proteins have a structural feature in common, they are different in regions outside the LRR. Analysis of the deduced amino acid sequences suggests that RPS2 and N' are cytoplasmic, whereas Cf-9 is extracellular, with a C-terminal membrane anchor, and L^6 is intracellular but attached to a membrane via an N-terminal signal anchor. Sequence analysis of RPS2 and N' also suggests that they have domains that may interact with other proteins.

The identification of plant resistance proteins predicted to have features such as kinase activity and protein–protein interactions, which are found in the proteins of signal transduction pathways, suggests a role for the proteins in specific pathogen recognition and induction of the HR. The molecular basis of plant–pathogen recognition and the expression of disease resistance is still not understood, but the recent cloning of several plant resistance and pathogen avirulence genes has provided important tools for further investigations.

14.5 Other plant resistance genes

Not all plant resistance genes follow the pattern of interactions shown in Figure 14.1 and *mlo* in barley is an example of such a gene. The *mlo* gene is situated on chromosome 4 and was originally described as a mutagen-induced barley mutant that was resistant to powdery mildew (*Erysiphe graminis* f. sp. *hordei*) but it has subsequently also been identified as a spontaneous mutant. All of the independently isolated *mlo* mutations map as non-complementing recessive alleles, that is they are mutations of the same gene. All of the *mlo* alleles confer the same resistance reaction and are effective against all races of powdery mildew. Leaves of resistant barley show virtually no sign of infection, that is there is no HR. Although all recessive *mlo* alleles are effective, there is some variation between individual *mlo* alleles and individual barley genotypes in the amount of background infection.

The *mlo* gene has not been cloned but it is thought to affect the formation of papillae in infected epidermal cells. Papillae are cell wall appositions that are probably formed as a general wound-healing mechanism. Powdery mildew only invades the epidermal cells and the *mlo* mutations do not affect the virulence of fungal pathogens, like rusts, which enter leaf stomata rather than the epidermis.

14.6 Resistance of plants to pests: cyanogenesis

A wide variety of mechanisms of resistance to pests exist in plants. In this section cyanogenesis will be described as an example that involves the formation of a secondary plant compound; the process has been studied at both the molecular and the genetic level.

The term 'cyanogenesis' describes the release of hydrocyanic acid (HCN) that occurs when plant tissue is damaged. It has been found to occur in over 2000 plant species, including the crops barley, flax, sorghum, cassava and white clover. HCN is formed when cyanoglucosides stored in the plant are sequentially broken down by two enzymes, a cyanogenic β-glucosidase and an α-hydroxnitrilase, to produce glucose, a ketone and HCN. In general there is no turnover of cyanoglucosides in plant tissues, with the cyanoglucoside and degrading enzymes being compartmentalized in the intact plant tissue. Typically cyanoglucosides are located in vacuoles and the first enzyme involved in degradation (the β-glucosidase) is extracellular, although in some plants, such as cassava, the cyanogenic β-glucosidase is located in specialized cells, the latex vessels.

Several of the genes involved in cyanogenesis have been cloned including the cyanogenic β-glucosidase from white clover and cassava, the α-hydroxynitrilase from cassava, black cherry and sorghum, and one of the enzymes involved in cyanoglucoside biosythesis from sorghum.

A generalized biosynthetic pathway for cyanogenesis is shown in Figure 14.2. The primary precursors of cyanoglucosides are the five hydrophobic protein amino acids valine, leucine, isoleucine, phenylalanine and tyrosine and one non-protein amino acid, cyclopentenylglycine. The cyanoglucoside biosynthetic pathway has proved difficult to elucidate because all but the final glycosylation step are catalysed by membrane-bound proteins. Recently it has been shown that in sorghum a single multifunctional cytochrome P450 catalyses the conversion of L-tyrosine into p-hydroxyphenyl-acetaldehyde oxime. This reaction involves four transition intermediate compounds. The activity of this remarkable protein was confirmed by the isolation of a cytochrome $P450_{TYR}$ cDNA clone from sorghum and the demonstration that the enzyme produced by an expression vector in *E. coli* was able to convert tyrosine into p-hydroxyphenyl-acetaldehyde oxime. The enzyme (or enzymes) converting the oxime into p-hydroxymandelonitrile has yet to be isolated from any plant. The final step in cyanoglucoside synthesis is the glycosylation of a hydroxynitrile by UDP-glucose glucosyltransferase to produce a stable compound. This biosynthetic pathway has long been recognized as a highly organized enzyme system that efficiently channels the intermediates between amino acid and cyanoglucoside.

Molecular and biochemical studies of cyanogenic β-glucosidases show that they have considerable sequence homology. This is a common feature of proteins having the same (or similar) biochemical function isolated from different plant species. However, the hydroxynitrilase enzymes that have been cloned from black cherry, sorghum and cassava are all different (Table 14.7) with essentially no sequence homology.

The best experimental evidence for the role of cyanogenesis in resistance to predation by small pests comes from the studies in white clover. White clover is polymorphic for cyanogenesis, that is natural populations and cultivars contain both cyanogenic and acyanogenic plants. This polymorphism is controlled by the

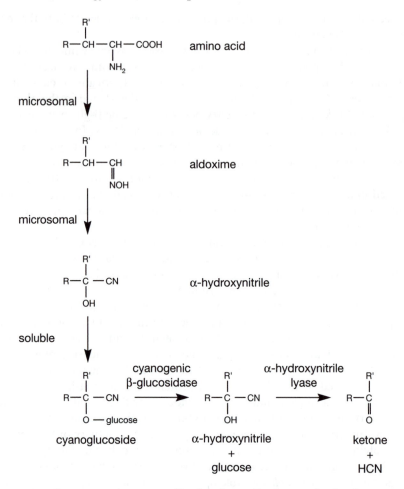

Figure 14.2 Cyanogenesis: a generalized pathway for the synthesis of cyanogenic glucosides and their degradation to release HCN. In cassava and white clover the cyanogenic glucoside, linamarin, has valine as a precursor, where R = CH$_3$ and R' = CH$_3$.

Table 14.7 A comparison of α-hydroxynitrile lyase from cassava (*Manihot esculenta* Crantz), sorghum (*Sorghum bicolor* L.) and black cherry (*Prunus serotina* L.)

Species	Cyanoglucoside	Precursor amino acid	Hydroxynitrile lyase		
			Glyco-protein	Flavin group	Subunit M_r
Cassava	Linamarin	Valine	No	No	28 500 (homotrimer)
Sorghum	Dhurrin	Tyrosine	Yes	No	22 000 + 33 000 (heterotetramer)
Black cherry	Prunasin	Phenylalanine	Yes	Yes	~59 000 (monomer)

alleles of two independently segregating genes (*Li* and *Ac*). The gene *Ac* controls the biosynthesis of cyanoglucosides and plants homozygous for the recessive non-functional allele of *Ac* appear to have all the cyanoglucoside biosynthetic pathway missing. The *Li* gene controls the presence or absence of the cyanogenic β-glucosidase and plants homozygous for the recessive non-functional allele of *Li* produce no cyanogenic β-glucosidase mRNA. The nature of the *Ac* and *Li* genes is not known; however the available evidence suggests that these genes may not be structural genes for the enzymes involved in cyanogenesis. White clover produces no α-hydroxynitrilase and it is assumed that the α-hydroxynitrile spontaneously decomposes to a ketone and HCN owing to its instability at high pH in the disrupted tissue.

Only plants containing at least one functional allele of both *Ac* and *Li* are cyanogenic. Because these two genes segregate independently, intercrossing heterozygous F_1 plants produces a classic 9:7 (cyanogenic:acyanogenic) modified Mendelian dihybrid segregation ratio in the F_2 generation (Table 14.8).

The cyanogenic polymorphism in white clover is maintained by two opposing selective forces. In cold environments there is selection **against** the cyanogenic phenotype, which is assumed to be due to increased frost damage in the cyanogenic morphs caused by HCN production following initial freezing damage. This selection is balanced by selection **for** cyanogenic phenotypes by small predators, such as slugs, causing greater damage to the acyanogenic morph. Feeding experiments have shown that some, but not all, species of slug and snail will discriminate between cyanogenic and acyanogenic morphs of white clover. In other words, the production of HCN protects white clover from predation by some species of mollusc.

Not all cyanogenic species are polymorphic for the character and the tropical root crop cassava (*Manihot esculenta* Crantz) is an example of this type of plant. In cassava there is variation between different plants in the levels of HCN produced. However, the only well-documented report of a differential pest attack on the low versus the high HCN plants is for the burrowing bug (*Cyrtomenus bergi*) in South America. Cassava evolved in South America and it is significant that the burrowing bug is an introduced species first recorded in Colombia in 1979. There is no evidence for the protective role of HCN against endogenous South American species since most of these appear to feed equally on the high and low HCN plants. This is believed to be a consequence of the co-evolution of cyanogenesis in the cassava plant and detoxifying systems in the potential pests.

Major Learning Objectives for Chapter 14

1. Knowledge of the biochemical and physiological bases of a variety of plant disease resistance mechanisms.
2. Understand the pattern of inheritance of plant resistance and pathogen virulence seen in gene-for-gene incompatible plant–pathogen interactions.
3. Distinguish between the molecular, genetic and biochemical bases of incompatible and compatible plant–pathogen interactions.
4. Knowledge of the cloned plant disease resistance genes.
5. Understand the ecological genetics of the cyanogenic polymorphism in white clover.

Table 14.8 Independent segregation of alleles of the *Ac* and *Li* genes that control cyanogenesis in white clover (*Trifolium repens* L.): production of 9:7 ratio in F_2 generation. HCN signifies the cyanogenic phenotype

$$F_1 \quad \frac{Ac \;\; Li}{ac \;\; li} \quad \times \quad \frac{Ac \;\; Li}{ac \;\; li} \quad F_1$$

$$\text{HCN} \qquad\qquad \text{HCN}$$

↓

F$_2$ generation

Gametes	Ac Li	Ac li	ac Li	ac li
Gametes				
Ac Li	$\frac{Ac \;\; Li}{Ac \;\; Li}$ HCN	$\frac{Ac \;\; li}{Ac \;\; Li}$ HCN	$\frac{ac \;\; Li}{Ac \;\; Li}$ HCN	$\frac{ac \;\; li}{Ac \;\; Li}$ HCN
Ac li	$\frac{Ac \;\; Li}{Ac \;\; li}$ HCN	$\frac{Ac \;\; li}{Ac \;\; li}$	$\frac{ac \;\; Li}{Ac \;\; li}$ HCN	$\frac{ac \;\; li}{Ac \;\; li}$
ac Li	$\frac{Ac \;\; Li}{ac \;\; Li}$ HCN	$\frac{Ac \;\; li}{ac \;\; Li}$ HCN	$\frac{ac \;\; li}{ac \;\; Li}$	$\frac{ac \;\; li}{ac \;\; Li}$
ac li	$\frac{Ac \;\; Li}{ac \;\; li}$ HCN	$\frac{Ac \;\; li}{ac \;\; li}$	$\frac{ac \;\; Li}{ac \;\; li}$	$\frac{ac \;\; li}{ac \;\; li}$

Further reading

BRIGGS, S.P. and JOHAL , G.S. (1994). Genetic patterns of plant host–parasite interactions. *Trends in Genetics*, **10**, 12–16.
A short review that focuses on the distinction between compatible and incompatible types of host–parasite interactions.

HUGHES, M.A. (1991). The cyanogenic polymorphism in *Trifolium repens* L. (white clover). *Heredity*, **66**, 105–115.
A review of the biochemistry and genetics of cyanogenesis together with an assessment of the evidence for a defence function of cyanogenesis.

KEEN, N.T. (1992). The molecular biology of disease resistance. *Plant Molecular Biology*, **19**, 109–122.
This review gives more information about mechanisms of disease resistance and less information about cloning than Staskawicz *et al.* (1995).

SCHÄFER, W. (1994). Molecular mechanisms of fungal pathogenicity to plants. *Annual Review of Phytopathology*, **32**, 461–477.
As the title suggests this review deals with plant disease primarily from studies of fungal pathogenicity.

STASKAWICZ, B.J., AUSUBEL, F.M., BAKER, B.J., ELLIS, J.G. and JONES, J.D.G. (1995). Molecular genetics of plant disease resistance. *Science*, **268**, 661–667.
A clear review that deals with the analysis of molecular clones of plant disease resistance genes, together with a section on evolution of disease resistance and a section on bioengineering disease resistance.

Part four

An introduction to plant biotechnology

Chapter 15

Transgenic plants

15.1 Plant biotechnology: commercial status

In the final part of this book the applications of molecular techniques to crop improvement (sometimes called genetic engineering) will be outlined. This chapter will deal with the improvement of agronomic characters, such as herbicide resistance, disease resistance and pest resistance and the improvement of quality, which is a character of yield. In the next chapter the modification of plants to produce novel materials will be covered together with a discussion of the concerns that have been expressed about the impact of plant biotechnology in areas such as food safety, the environment and the developing world.

The Green Industry Biotechnology Platform is an industry association of 21 European companies actively using biotechnology for plant improvement. The database produced by this association shows that between 1989 and 1993 nearly 200 different properties genetically engineered into 38 crop species have undergone field trials (Table 15.1). Over half of all the tests were carried out in North America. This organization estimates that by the year 2000 over 400 plant–character combinations will be under field trial. Although only a few genetically engineered crops are currently in the market place, this list of field trials indicates the number of products in the pipeline.

Table 15.2 shows the number of field evaluations registered for the five most commonly modified crops, broken down by the trait (character) that has been manipulated. Potato is involved in more trials than any other crop, followed by oilseed rape, tobacco, maize and tomato. The large number of trials of modified maize are surprising given the fact that maize is a monocotyledon and cannot be transformed with the *Agrobacterium* Ti-based system (see section 15.2). Clearly the level of investment in maize, indicated by the numbers in Table 15.2, reflects the economic importance of this crop. Ranking the traits or characters shows that herbicide resistance is the character manipulated more often than any other trait. Herbicide resistance is followed by quality improvement, virus resistance and insect or pest resistance. Excluding environmental trials (marker genes), these four characters make up 90% of all the projects. Despite the domination of plant biotechnology by these four traits, a very wide

Table 15.1 Crop plants able to be transformed and regenerated

Cereals	Fibre crops	Legumes and oil seeds	Horticultural crops	Pasture crops	Trees	Tropical crops
Rice	Cotton	Bean[a]	Aubergine, asparagus, cabbage, carrot cauliflower, celery[a], chicory, chrysanthemum, cucumber, gerbera, kiwi fruit, lettuce, melon, petunia, potato, service berry, squash, sugarbeet, tobacco, tomato	Alfalfa	Apple	Banana[a]
Maize		Flax		Orchard grass[a]	Birch	Papaya
Wheat		Oilseed rape		White clover[a]	Eucalyptus	Sugar cane
Barley[a]		Pea[a]			Poplar	
Rye[a]		Peanut			Plum	
		Soybean			Spruce	
		Sunflower			Walnut	

[a]Crops not undergoing field trials (reported up to 1994).

Table 15.2 Number of field trials with the top five genetically modified plants (1986–1993)

Trait	Potato	Oilseed rape	Tobacco	Maize	Tomato	Total
Herbicide resistance	16	94	29	54	21	214
Quality improvement	31	57	13	15	39	155
Virus resistance	60	2	24	10	20	116
Insect resistance	34	3	19	24	16	96
Marker gene (used to monitor environment impact)	23	17	28	8	4	80
Fungal resistance	9	5	9	2	0	25
Multiple traits	8	2	4	0	0	14
Bacterial resistance	9	1	0	0	0	10
Unspecified	3	1	5	5	3	17
Total	193	181	128	120	105	727

range of different genes from a variety of different organisms are currently being introduced into plants, and examples of these will be described in Chapter 16.

15.2 Alternative methods for plant transformation

There are two distinct stages to the successful transformation of fertile plants: (i) the stable integration and expression of foreign DNA into nuclear chromosomes and (ii) the growth and regeneration of a normal flowering plant from transformed cells. The use of *A. tumefaciens* Ti plasmid has been described in Chapters 6 and 7. However, *A. tumefaciens* is only a disease of dicotyledons and it cannot be used to transform monocotyledon cells. In addition, *A. tumefaciens* Ti-based transformation does not work equally well on all dicotyledonary species and in many dicotyledon and monocotyledon species the lack of good reliable tissue culture and regeneration protocols also limits transformation technology. In view of the economic importance of the monocotyledon cereals, a great deal of money and effort is being spent to develop reliable transformation and regeneration protocols for these plants.

A variety of different techniques have been explored as alternatives to Ti plasmid transformation, including electroporation into protoplasts (plant cells with the cell wall removed enzymatically), viral vectors, agroinfection using a combination of viral and T-DNA, microinjection and particle gun delivery systems or biolistics. It is biolistics that has been the most successful of these alternative techniques and is the method that was used to transform the cereals shown in Table 15.1. Although there are reports of transformation in wheat, barley and rye, difficulties with the culture and regeneration of these crops means that a cultivar-independent biolistic transformation protocol does not exist for these crops. However maize and rice have been transformed with a variety of genes (Table 15.2 for maize).

The particle delivery system or biolistic method is conceptually very simple. Small particles of gold or tungsten are coated with the DNA construct and shot into the plant cells. In practice the procedure is not completely straight-forward. The problem of delivering reproducible quantities of DNA into cells without damaging them, so that they can regenerate, has led to the development of a number of types of equipment. One very successful method is to load the coated particles on to a membrane that is projected from its holder by a controlled pressure of helium gas. The membrane plus particles hits a stop mesh, which retains the membrane but allows the accelerated particles to pass through into the plant tissue below (Bio-Rad, Biolistic PDA-1000/He). The method can be optimized for different tissues by altering the gas pressure, the particle size and the distance the particles travel. Other particle bombardment systems use an electric discharge to accelerate the particles and this has been used successfully with rice and soybean. Biolistics has also been used to deliver DNA into chloroplasts, allowing integration of foreign DNA into the chloroplast genome by homologous recombination (see section 15.6).

15.3 Engineering herbicide resistance in plants

Engineering herbicide resistance in plants has received a lot of effort and money for two main reasons. Firstly, herbicide resistance in crop plants is of considerable importance and potential commercial benefit to chemical companies who produce herbicides. Secondly, studies of mutant herbicide-resistant weeds indicate that this character can be due to a dominant mutation of a single gene, which means that only one gene needs to be manipulated to produce a transgenic plant with herbicide resistance. The technique is also facilitated by the powerful selection for transgenic plants made possible by their ability to grow in the presence of the herbicide.

An ideal herbicide should: (i) not kill animals, (ii) be biodegradable, and (iii) kill weeds and not crops. It is the last of these criteria that is the biggest problem since, in general, herbicides affect biochemical functions specific to plants, such as photosynthesis or essential amino acid biosynthesis, but these are reactions common to all plants. Although there is some known and exploited variation between plants in resistance to some herbicides, current selectivity usually comes from agronomic practice, such as the time or the site of application of the chemical.

Table 15.3 shows the biochemical basis of resistance to eight types of herbicide where (with one exception) the target protein for herbicide action is known. Most of the herbicides in this list affect either amino acid biosynthesis or photosynthesis and, with the possible exception of 2,4-D, they all affect proteins located in chloroplasts. Prokaryote species that are herbicide resistant have been isolated and those which have been used as the source of genes for engineering herbicide-resistant plants are also included in Table 15.3.

Two basic strategies have been used in the successful production of transgenic herbicide-resistant plants: (i) adding a mutant gene that encodes a

Table 15.3 Properties of herbicides and herbicide-resistance mechanisms in plants and microorganisms

Pathway	Herbicide	Target protein	Basis of resistance in mutant organisms or biotypes and cultivars
Amino acid biosynthesis	Glyphosate (non-selective)	5-Enolpyruvyl shikamate-3-phosphate synthase (EPSPS)	No plants. *Salmonella typhimurium, Escherichia coli, Agrobacterium tumefaciens* mutant EPSPS
	Sulphonylurea, imidazolinone (selective)	Acetolactate synthase (ALS)	*Arabidopsis* plus weed species, single substitution of proline 194 in ALS; detoxification in maize and wheat by hydroxylation and conjugation
	Phosphinothricin (non-selective)	Glutamine synthase (GS)	No plants. *Streptomyces hygroscopicus* detoxification by acetylation
Photosynthesis: photosystem II	Triazine (selective)	Q_B-binding protein (D1)	Maize, detoxification by 2-hydroxylation and conjugation to glutathione; weed species, serine to glycine amino acid change in D1; *Abutilon theophrasti* (weed) detoxification by enhanced glutathione *S*-transferase (conjugating) activity
	Bromoxynil (non-selective)	Q_B-binding protein (D1)	*Alopercurus myosuroides* (weed), detoxification by cytochrome P450 monooxygenase; *Klebsiella ozaenae* detoxification by nitrilase
Photosynthesis: photosystem I	Paraquat (non-selective)	Generates superoxide radicals in light	Weeds, detoxification of reactive oxygen by superoxide dismutase
Lipid biosynthesis	Aryloxyphenoxy-propionate (selective)	Acetyl coenzyme A carboxylase (ACCase)	Dicotyledons resistant ACCase; wheat detoxification by esterification, hydroxylation and *O*-glycosylation
Phytohormone: auxin	2,4-D (selective)	Unknown	*Alcaligenes* spp. detoxification by cytochrome P450 monooxygenase; weed and crop (monocotyledons) mechanism unknown

target protein with altered sensitivity to the herbicide, and (ii) introducing a detoxifying enzyme.

Target proteins with altered sensitivity to the herbicide

Glyphosate

The herbicide glyphosate inhibits the enzyme 5-enolpyruvylshikimate-3-phosphate synthase (EPSPS) (Figure 15.1), which functions in the biosynthesis of essential aromatic amino acids (tryptophan, tyrosine, phenylalanine). The herbicide is active against all plant species (non-selective). It is widely used because it is effective against 76 of the world's 78 worst weeds and is rapidly degraded in the soil.

Glyphosate also inhibits aromatic amino acid synthesis in bacteria and mutant bacteria (*Salmonella typhimurium* and *E. coli*) have been selected that can grow in the presence of normally lethal glyphosate concentrations. These mutants contain an altered EPSPS that is not inhibited by glyphosate. The *E. coli* EPSPS gene (*AroA*) was converted into a plant nuclear gene that would produce a chloroplast-targeted enzyme by placing the normal *Arabidopsis* EPSPS transit peptide 5' of the bacterial *AroA* coding sequence and flanking this construct with a plant nuclear promoter and terminator sequence. Transgenic soybean plants with full resistance to twice the normal glyphosate application have been produced and undergone field trials.

There is a problem associated with the mutant *E. coli* EPSPS enzyme: the mutation causing reduced affinity for the herbicide (increased K_i) also reduces the enzyme's affinity for the substrate, phosphoenolpyruvate (increased K_m). More recently, an *Agrobacterium* EPSPS gene (*CP4*) has been identified that is resistant to glyphosate inhibition but has a low K_m for phosphoenolpyruvate and this has been used to produce transgenic oilseed rape (canola) and soybean. Glyphosate-resistant oilseed rape has been field tested annually since 1987 as part of a continuous research programme.

Sulphonylurea

Sulphonylureas are broad-spectrum herbicides with very low toxicity to animals. Maize and wheat are naturally resistant because they can detoxify this

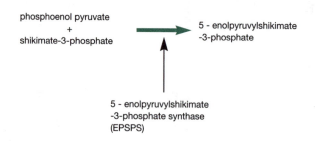

phosphoenol pyruvate
+
shikimate-3-phosphate

5 - enolpyruvylshikimate
-3-phosphate

5 - enolpyruvylshikimate
-3-phosphate synthase
(EPSPS)

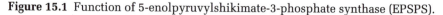

Figure 15.1 Function of 5-enolpyruvylshikimate-3-phosphate synthase (EPSPS).

herbicide (Table 15.3). Mutant *Arabidopsis* plants resistant to sulphonylurea have been isolated; these have a single base mutation in the gene encoding acetolactate synthase (ALS), the enzyme inhibited by sulphonylurea. This enzyme is part of the biosynthetic pathway of the essential amino acids valine, leucine and isoleucine. The *Arabidopsis* ALS mutation causes a replacement amino acid at proline-194 and the same mutational change has been shown to be the basis of sulphonylurea resistance that has arisen in many weeds. Transgenic tobacco plants have been produced expressing the *Arabidopsis* mutant ALS gene and these plants can tolerate a herbicide concentration four times a normal field application. Extensive field trials of transgenic flax, containing the *Arabidopsis* mutant ALS gene, show that the plants have good resistance to sulphonylurea under field conditions and yield as well as the parent cultivar.

Triazine

This group of herbicides kill plants by disrupting photosynthetic electron transport in photosystem II. They compete with plastoquinone for the Q_B-binding protein (D1), which is embedded in chloroplast thylakoid membranes. This protein is encoded by a chloroplast gene, *psb*A; triazine-resistant weeds have been shown to have mutations of this gene causing a single amino acid substitution in D1 (Table 15.3). Since *psb*A is a chloroplast gene and transformation technology manipulates nuclear genes, the transformation strategy converted a mutant *psb*A gene from *Amaranthus hybridus* into a nuclear gene by adding the 5' chloroplast transit peptide sequence from the ssRubisco plus a nuclear promoter and termination sequence to the mutant *psb*A coding sequence. Although the chimeric gene confers triazine resistance to transgenic tobacco there are problems with this strategy, which arise from the fact that the *psb*A mutation also limits the rate of electron transport in photosystem II. This affects photosynthesis and therefore plant growth and crop yield.

Introduction of a herbicide-detoxifying enzyme

Because the mutant target enzyme, with altered sensitivity for the herbicide, often has altered function (see EPSPS and D1) this method of producing herbicide resistance has inherent problems. In addition, of course, the transgenic plant will produce a mixture of proteins, partly from the mutant transgene but also from the normal endogenous genes. These considerations mean that the second approach, of adding a detoxification mechanism, is preferred. However, the major setback to this approach is the limitation of known plant detoxification systems that have a simple biochemical basis. Many of the best characterized plant detoxification systems have at least two stages, a hydroxylation and a conjugation stage (Table 15.3). However, the powerful method of selection for growth in the presence of a lethal concentration of herbicide has led to isolation of a number of prokaryote detoxifying enzymes and these have been used successfully to produce herbicide-resistant plants. In each case the prokaryotic coding sequence is used to produce a chimeric gene with plant nuclear signals for transcription (promoter plus terminator sequence). It has not been necessary to target these enzymes into the chloroplasts.

Phosphinothricin

Phosphinothricin acts on amino acid synthesis and inhibits glutamine synthetase (Table 15.3). The *bar* gene from *Streptomyces hygroscopicus* encodes phosphinothricin acetyltransferase, which acetylates phosphinothricin to a non-toxic compound. Transgenic plants expressing this gene are fully resistant to the herbicide.

Bromoxynil

Bromoxynil affects the D1 protein of photosystem II (Table 15.3). A strain of *Klebsiella ozaenae* was selected by its ability to use the herbicide bromoxynil as its sole nitrogen source. This strain of *Klebsiella* was able to degrade bromoxynil using a nitrilase, which is encoded by a plasmid gene, *bxn* (Figure 15.2). This gene was used to transform tobacco and tomato and bromoxynil-resistant transgenic plants were selected. Figure 15.2 shows genetic studies on three individual transgenic tobacco plants, which were either self-fertilized or crossed

(a)

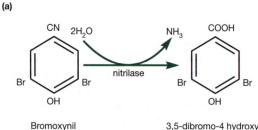

Bromoxynil

3,5-dibromo-4 hydroxy benzoic acid (non-toxic)

(b)

Cross	Analysis: number of progeny		
	Resistant	Sensitive	Ratio
39-3 self	102	48	3:1
39-3 × Xanthi	106	98	1:1
40-1 self	189	11	15:1
40-1 × Xanthi	170	34	3:1
40-6 self	160	47	3:1
40-6 × Xanthi	103	95	1:1

Figure 15.2 (a) Detoxification of bromoxynil by nitrilase and (b) segregation analysis of nitrilase activity (bromoxynil resistance) in the progeny of three transgenic *Nicotiana tabacum* plants.

with the parent (non-transgenic) cultivar Xanthi. Since the primary transformant will have the transgene in only one copy of the pair of homologous chromosomes, it will be heterozygous and self-fertilization will produce herbicide-resistant and herbicide-sensitive progeny in a 3:1 ratio (the transgene is dominant). Crossing with the non-transgene parent (Xanthi) will give a 1:1 ratio. Figure 15.2 shows that plants 39-3 and 40-6 behave like this, indicating that a single copy of the *bxn* gene is incorporated into the tobacco genome. However, the plant 40-1 gives a 15:1 ratio on self-fertilization and a 3:1 ratio when crossed with Xanthi. These are modified Mendelian dihybrid segregation ratios and indicate that plant 40-1 has two copies of the transgene, which are not linked, in its genome.

2,4-D

The target protein for 2,4-D and the mechanism of resistance in plants are unknown (Table 15.3); however some soil microorganisms can degrade 2,4-D. *Alcaligenes* spp. contain a plasmid-encoded cytochrome P450 monooxygenase that can degrade 2,4-D. The gene for this enzyme has been cloned and used to produce 2,4-D-resistant tobacco and cotton plants.

A summary of genetically engineered herbicide-resistant plants is given in Table 15.4.

15.4 Improving quality characters by plant transformation

The important limiting quality characters will be quite different for each particular crop and this is therefore a very heterogeneous trait. A few examples will be described in order to illustrate the breadth of projects in progress:

1. delayed softening of tomato fruits;
2. production of laurate in oilseed rape (canola);
3. increase in tomato fruit solids;
4. high lysine soybean.

All of these examples involve different genes and a different strategy but the success of each is dependent upon a knowledge of the biochemistry and developmental control of the process being manipulated. The first two examples represent products in commercial production and the last two examples are at pre-commercial development.

Delayed softening of tomato fruits

The Flavr Savr tomato produced by Calgene (Davis, California, USA) has been transformed with the antisense construct for polygalacturonase (see section 7.4). This enzyme is synthesized *de novo* during ripening and plays a role

Table 15.4 Genetically engineered herbicide-resistant plants

Transgenic plants	Herbicide[a]	Strategy
Changing properties of target protein		
Tobacco, tomato, soybean	Glyphosate (EPSPS)	Mutant *AroA* gene (EPSPS) from *Escherichia coli* or *Salmonella typhimurium*, plus 5' chloroplast transit peptide sequence from *Arabidopsis* EPSPS[b]
Oilseed rape (canola), soybean	Glyphosate (EPSPS)	*CP4* EPSPS from *Agrobacterium* plus 5' chloroplast transit peptide sequence from *Arabidopsis* EPSPS
Tobacco	Sulphonylurea (ALS)	Mutant ALS gene for *Arabidopsis*
Tobacco	Triazine (D1)	Mutant chloroplast D1 (*psb*A) gene from *Amaranthus hybridus* plus 5' chloroplast transit peptide sequence from ssRubisco[c]
Introduction of detoxifying enzyme		
Tobacco, tomato, potato	Phosphinothricin (GS)	*bar* (acetyltransferase) gene from *Streptomyces hygroscopicus*
Tobacco, cotton	Bromoxynil (D1)	*bnx* (nitrilase) gene from *Klebsiella ozaenae*
Cotton	2,4-D (?)	Cytochrome P450 monooxygenase gene from *Alcaligenes*

[a] Target protein given in parentheses (see Table 15.3).
[b] Increase in K_1 for glyphosate associated with increased K_m for substrate.
[c] Mutation limits rate of electron transport, therefore affects chloroplast function, growth and production.

in the depolymerization and solubilization of the pectin fraction of the fruit (pericarp) cell walls. The polygalacturonase antisense transgenic tomato has a very reduced level of enzyme activity, so the tomato fruits do not soften as quickly as normal tomatoes and therefore have an extended shelf-life. Fresh Flavr Savr tomatoes are on sale in the USA and are being used to produce soup and ketchup. A similar transgenic tomato cultivar called Endless Summer has been produced by DNA Plant Technology (Oakland, California, USA) using a similar strategy.

Production of laurate in oilseed rape (canola)

The same company that produced the Flavr Savr tomato (Calgene, Davis, California, USA) has produced a transgenic oilseed rape (canola) that synthesizes significant quantities of the medium-chain fatty acid lauric acid ($C_{12:0}$), which is used in the production of soap. The oil from normal canola has very little lauric acid and is 62% oleic acid ($\Delta 9C_{18:1}$). The transgenic oilseed rape contains a gene for acyl-ACP thioesterase from the California bay tree (*Umbellularia californica*), which produces up to 70% lauric acid in its seed triacylglycerols. Acyl-ACP thioesterases play a central role in fatty acid synthesis because they determine the chain lengths of the fatty acids transported to the cytosol. The enzyme hydrolyses fatty acyl-ACP (acyl carrier protein) releasing ACP and the fatty acid. Transgenic oilseed rape plants have 45% of their seed oil as lauric acid ($C_{12:0}$). The plants have a normal appearance, and produce normal yields of seed oil. This genetically engineered cultivar is grown commercially in southern Georgia, USA, and the oil is currently on sale in the USA.

Increase in tomato fruit solids

A reduction in the amount of water in tomatoes and a concomitant increase in solid compounds is of considerable economic importance in tomato processing. It is also important in fresh fruit since sugar and organic acids, which make up most of the solid compounds in a tomato fruit, are important components of flavour. It has been shown that the plant hormone cytokinin influences the transport and deposition of photosynthates in young reproductive tissues. Isopentenyl AMP is a precursor of cytokinin and is synthesized from isopentenyl pyrophosphate and AMP by a reaction that is catalysed by the enzyme isopentenyl transferase. A chimeric isopentenyl transferase gene was constructed using a promoter from an ovary-specific tomato gene and transgenic tomatoes were produced that expressed the isopentenyl transferase at high levels in developing ovaries. Cytokinin levels were elevated in the ovaries of transgenic plants with soluble solids significantly increased in field trials of five of the seven independent transgenic tomato lines.

High lysine soybean

Lysine is an essential amino acid that is limiting in most cereal crops including maize. For animal feed, cereal grain has to be supplemented annually by 200 000 tons of crystalline lysine produced by microbial fermentation. One approach to improving the lysine content of seeds is to increase the amount of free lysine (rather than lysine found as a component of the storage proteins; see Chapter 9). In plants, lysine is derived from aspartate (Figure 15.3) and its synthesis is controlled by feedback inhibition. Two enzymes in the lysine biosynthetic pathway are inhibited by lysine and these serve as regulatory points. The first

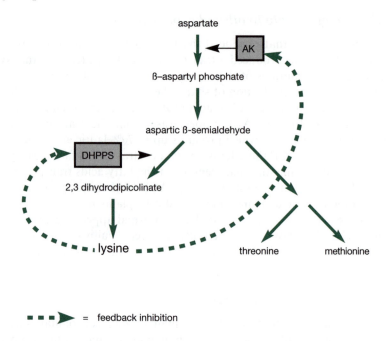

aspartate

AK

β–aspartyl phosphate

aspartic β-semialdehyde

DHPPS

2,3 dihydrodipicolinate

lysine   threonine   methionine

◼ ◼ ◼▶   =   feedback inhibition

Figure 15.3 Lysine biosynthetic pathway showing feedback inhibition of aspartokinase (AK) and dihydrodipicolinic acid synthase (DHDPS).

control-point enzyme is aspartokinase (AK), which phosphorylates aspartate; the second enzyme is dihydrodipicolinic acid synthase (DHDPS), which carries out a condensation reaction of aspartic β-semialdehyde and pyruvate. This is the first committed step in the lysine pathway.

The bacterial DHDPS is much less sensitive to inhibition by lysine than the plant enzyme and the *Corynbacterium* DHDPS gene *dap*A has been cloned. A mutant *E.coli lys*C gene, which encodes a lysine-insensitive AK, has also been cloned. Each of these genes was linked to a chloroplast transit peptide and a seed-specific promoter and used to produce a transgenic soybean that expressed both enzymes. The transgenic plants had a several hundred-fold increase in seed free lysine and the total seed lysine content was increased five-fold. However some increase in intermediates in lysine biosynthesis was observed and the nutritional consequence of this has to be evaluated.

15.5 Engineering plants for virus resistance

Plant virologists reported the phenomenon of cross-protection of plants as early as 1929. The term 'cross-protection' describes the phenomenon that occurs when a plant is inoculated with a mild strain of virus (causing minimal disease symptoms): this infection produces partial protection against a subsequent inoculation with a virulent viral strain. In many cases this cross-protection is related to the synthesis of the virus coat protein within the infected plant cells.

In 1986 Roger Beachy's research group showed that transgenic tobacco plants expressing the tobacco mosaic virus (TMV) coat protein were protected against TMV infection. Since that time protection against viral infection in viral coat protein transgenic plants has been found for a number of viruses, including alfalfa mosaic virus, tobacco streak virus, tobacco rattle virus, cucumber mosaic virus, soybean mosaic virus, potato virus X, potato virus S, potato leaf roll virus, tomato spotted wilt virus and potyviruses, such as maize dwarf mosaic virus. Both classical cross-protection and viral coat protein transgene protection is less effective when viral RNA (rather than virus particles) is used as the infectious agent. However, the precise mechanism of protection is still not known.

Seeds of transgenic yellow crookneck squash with viral coat protein protection against water melon mosaic virus 2 and zucchini yellow mosaic virus are on sale in the USA. This technique is a powerful tool for improving crops, particularly those like potato, which are vegetatively propagated, where viral diseases can be transmitted from year to year through the vegetative planting material (NB potato is also a tetraploid species, which makes conventional breeding more difficult than many crops).

15.6 Engineering plants for insect resistance

The bacterium *Bacillus thuringiensis* (*Bt*) is subdivided into a number of subspecies and strains that are distributed all over the world. When these bacteria form spores they produce a large crystal-like structure made out of proteins. One of these crystal proteins (Cry) has potent insecticidal activity. The protein is often called the *Bt* toxin. The protein exists as a protoxin (about $130-140 \times 10^3 \, M_r$) that is proteolytically cleaved in the insect larva midgut when the bacterial spores are eaten. Proteolysis produces the active toxin ($60-70 \times 10^3 \, M_r$), which binds to the membrane of cells that line the gut and permeabilizes them causing a breakdown of cell function, starvation and death of the larva.

There has been co-evolution of bacterium and insects so that different strains of bacterium produce different Cry proteins which will only kill a limited number of insect species. The protoxin proteins are classified CryI to CryV by amino acid sequence homology and are known to be toxic to Lepidoptera (moths and butterflies), Diptera (flies) and Coleoptera (beetles) but not to other animals.

The *cry* (*Bt* toxin) genes from a number of *Bt* strains have been cloned and expressed in transgenic plants. However, the genes are A/T-rich compared to plant genes and expression in plants is limited by the presence of plant nuclear processing signals in the coding sequence and by the codon usage of nuclear genes (see section 7.4). Synthetic genes have therefore been constructed in which the coding region has been transformed into a plant-like coding sequence. Despite this problem and the problem of specificity of action, a number of *Bt* toxin transgenic plants have been produced and, after undergoing extensive field trials, are now approved for commercial release:

Cotton	*Bt* toxin against cotton boll worm	Monsanto, St Louis, Missouri, USA
Potato	*Bt* toxin against Colorado beetle	Mycogen, San Diego, California, USA
Maize	*Bt* toxin against European corn borer	Ciba Seeds, Greensboro, North Carolina, USA

The recoding of *Bt cry* (*Bt* toxin) genes in order to ensure levels of expression that will be active against insect larvae is very expensive and laborious, particularly since they are large genes (approximately 3.5 kb). However, the transcriptional and translational machinery of chloroplasts is very similar to that of prokaryotes and in addition the chloroplast genome is relatively A/T-rich. Recently an unmodified *Bt kurstaki* HD 73 *cry*IA(c) gene has been stably incorporated into a chloroplast genome. Since plant cells may contain up to 50 000 copies of the chloroplast genome, transgene incorporation into a chloroplast genome producs automatic amplification of the gene in each cell. Tobacco plants containing the transgenic chloroplasts were shown to produce 3–5% of total soluble leaf protein as *Bt* protoxin and to be extremely toxic to the tobacco budworm and armyworm.

Box 15.1 Production of transgenic plants containing the transgene in all chloroplast genomes (homoplastic)

The wild-type bacterial *cry*IA(c) coding sequence was flanked by the ribosome-binding region of the ssRubisco chloroplast gene and the 3' UTR of the chloroplast *rps*16 ribosomal protein gene and attached to the strong, constitutive chloroplast ribosomal RNA operon promoter (*Prrn*). The vector used for transformation also contains flanking tobacco plastid DNA homology regions to direct insertion of the *Bt* transgene into the tobacco chloroplast genome between the *trn*V and *rps*12/7 loci by homologous recombination. The DNA was introduced into leaf chloroplasts on tungsten particles using biolistics. Transformed chloroplast genomes were selectively amplified by growing the cells on spectinomycin medium, which selects plastids carrying the *aad*A (spectinomycin resistance) gene, which is linked to the *cry*IA(c) gene in the transformation construct.

An alternative strategy for the production of insect-resistant plants has used naturally occurring plant defence mechanisms. Many plant defence mechanisms are based on physical barriers to being eaten (spines, hairs) or on the production of secondary plant compounds synthesized by strictly regulated biosynthetic pathways (see cyanogenesis, section 14.6). To introduce such systems using genetic engineering would involve the coordinated transfer of many genes and is outside the current scope of the technology. However, some plants produce defence proteins that are digestive-enzyme inhibitors or lectins.

Lines of cowpea (*Vigna unguiculata* L.(Walp.)) resistant to attacks by the brucid beetle, an important pest of the stored crop, were shown to contain high levels of trypsin inhibitor (CpT1). Transgenic tobacco plants containing the cowpea trypsin inhibitor gene under the control of a constitutive promoter were shown to have significantly enhanced resistance to a broad range of pests that were capable of eating tobacco. The gene for wheat germ agglutinin, a chitin-binding lectin that inhibits larval growth, is also under study as a potential pest resistance agent. The problem with lectins and digestive-enzyme inhibitors is that they can also affect humans and domestic animals. The potential hazard of these proteins could be contained if, for example, they were used selectively to control attack by root pests by placing the gene under the control of a root-specific promoter.

Major Learning Objectives for Chapter 15

1. Knowledge of the status of the commercial production of transgenic crops and products.
2. Knowledge of biolistic technology for plant transformation.
3. Knowledge of the successful examples of transgenic herbicide-resistant plants and appreciation of the difference between the two strategies that have been used.
4. Knowledge of the scope for improvement of crop quality characters using plant transformation technology.
5. Knowledge of strategies to produce virus- or insect-resistant transgenic crops.
6. Understand why there has been commercial investment in transformation technology for the production of crops with herbicide resistance, improved quality characters and virus and insect resistance, rather than more investment in conventional plant breeding for these traits.

Further reading

The following two reviews cover most aspects of the production of transgenic herbicide-resistant crop plants.

OXTOBY, E. and HUGHES, M.A. (1990). Engineering herbicide tolerance into crops. *Trends in Biotechnology*, **8**, 61–65.

HOLT, J.S., POWLES, S.B. and HOLTUM, J.A.M. (1993). Mechanisms and agronomic aspects of herbicide resistance. *Annual Review of Plant Physiology and Plant Molecular Biology*, **44**, 203–229.

AHL GOY, P. and DUESING, J.H. (1995). From pots to plots: genetically modified plants on trial. *Biotechnology*, **13**, 454–458.
This reference gives statistical information about the commercial implementation of transgenic plant technology.

The following two references cover the biotechnological opportunities that exist in lipid biosynthesis modification.

CHASAN, R. (1995). Engineering fatty acids – the long and the short of it. *Plant Cell*, **7**, 235–237.

TÖPFER, R., MARTINI, N. and SCHELL, J. (1995). Modification of plant lipid biosynthesis. *Science*, **268**, 681–685.

The following three references review the production of transgenic plants with either virus or insect resistance.

GATEHOUSE, A.M.R., BOULTER, D., and HILDER, V.A. (1992). Potential of plant-derived genes in the genetic manipulation of crops for insect resistance. In Gatehouse, A.M.R., Hilder, V.A. and Boulter, D. (eds) *Plant genetic manipulation for crop protection*, pp. 155–181. CAB International, Wallingford.

PEFEROEN, M. (1992). Engineering of insect-resistant plants with *Bacillus thuringiensis* crystal protein gene. In Gatehouse, A.M.R., Hilder, V.A. and Boulter, D. (eds) *Plant genetic manipulation for crop protection*, pp. 135–153. CAB International, Wallingford.

STURTEVANT, A.P. and BEACHY, R.N. (1993). Virus resistance in transgenic plants: coat protein-mediated resitance. In Hiatt, A. (ed.) *Transgenic plants*, pp. 93–114. Marcel Dekker, New York.

The September 1995 issue of *Trends in Biotechnology* (vol. 13, no. 9) is a special issue devoted to plant biotechnology and contains the following four articles.

FLAVELL, R.B. (1995). Plant biotechnology R&D – the next ten years. *Trends in Biotechnology*, **13**, 313–319.

MAZUR, B. (1995). Commercializing the products of plant biotechnology. *Trends in Biotechnology*, **13**, 319–323.

SHAH, D.M., ROMMEUS, C.M.T. AND BEACHY, R.M. (1995). Resistance to diseases and insects in transgenic plants: progress and applications to agriculture. *Trends in Biotechnology*, **13**, 362–368.

WALDEN, R. AND WINGENDER, R. (1995). Gene-transfer and plant regeneration techniques. *Trends in Biotechnology*, **13**, 324–331.

Chapter 16

Prospects for plant biotechnology

16.1 Plant biotechnology: future prospects and concerns

In Chapter 15 a number of examples of the use of molecular techniques to modify plants for agronomic and yield characters were described. These examples are largely near-market projects and are limited to a relatively small number of characters. In this chapter the use of molecular techniques to produce crop plants with novel characteristics will be introduced. This type of plant biotechnology, where plants are treated as bioreactors for the production of specific compounds, has been called molecular farming.

One of the major limitations in plant biotechnology is still our relatively poor understanding of fundamental plant processes. Nevertheless, it is considered an exciting field with vast prospects for crop improvement and for novel plant production systems (molecular farming). Despite the benefits of the low energy input and renewable resources such novel systems offer, there are concerns about the technology that are the subject of debate by the general public, the press, legislative bodies and scientists. This chapter will introduce the widespread debate about plant biotechnology and highlight some of the concerns and issues the use of this new technology raises.

16.2 Molecular farming

Table 16.1 lists nine compounds that have been successfully synthesized in transgenic plants. These compounds range from simple peptides to a thermoplastic, and the source of the genes used varies from bacteria to human cells. The list is not comprehensive but it illustrates both the great versatility of plant systems and the considerable ingenuity of plant molecular scientists. Since plant transformation technology was only developed in the early 1980s the application of this technology has developed extraordinarily rapidly.

There are broadly two types of product: (i) high-value compounds that would have relatively small-scale production requirements, such as pharmaceutical products; and (ii) compounds required on a bulk scale with low

Table 16.1 Examples of the construction of transgenic plants for the production of novel compounds

Plant species	Compound	Use	Origin of gene
Arabidopsis thaliana	Polyhydroxyalkanoates (polyhydroxybutyrate)	Biodegradable thermoplastic	*Alcaligenes eurtrophus* (bacterium)
Brassica napus (oilseed rape or canola)	Leu-enkephalin	Neuropeptide	Human
Brassica napus	Magainin	Antibacterial peptide	Frogs
Nicotiana tabacum (tobacco)	Thermostable xylanase	Biomass processing, paper industry	*Clostridium thermocellum* (bacterium)
Nicotiana tabacum	Catalytic antibodies	Cancer treatment	Human hybridoma cell line
Nicotiana tabacum	α-Amylase	Food processing	*Bacillus licheniformis* (bacterium)
Nicotiana tabacum infected with tobacco mosaic virus	Malaria; epitopes (AGDR, QGPGAP)	Malaria vaccine	Malarial parasite
Solanum tuberosum (potato)	Enterotoxin antigen	Cholera oral vaccine	*Vibrio cholerae* (bacterium)
Solanum tuberosum	Serum albumin	Broad clinical use	Human

production costs. One of each of these types is described below. It is considered that plant biotechnology has the strongest prospects in the production of bulk-scale low-cost products. In special circumstances plant biotechnology will be used for some high-value small-scale products.

Example 1. High-value pharmaceutical compound: malarial epitope vaccine

Malaria is a widespread and serious disease where the production of normal vaccines is difficult and where the use of molecular techniques offers great advantages over conventional techniques. In this example, the plant (tobacco) is used to manufacture genetically engineered virus particles. A continuous epitope is a linear peptide fragment of an antigen (protein) that binds to antibodies raised against the complete protein. Conversely, antibodies raised against an epitope can neutralize (bind to) the complete protein. However, the immunogenicity of these short peptide sequences is greatly enhanced if they are 'presented' to the immune system as part of a larger immunogenic molecule.

TMV is a rod-shaped RNA virus. It has been extensively studied and the pH-controlled *in vitro* polymerization of TMV coat protein into helical rods that are stable at room temperature has been known for some time. Genetically stable recombinant TMV viruses were constructed that contained either the malarial epitope AGDR in an internal loop or the malarial epitope QGPGAP as a C-terminal extension to the coat proteins. Since TMV is an RNA virus the genetic manipulation had to be carried out on the cDNA sequence (produced from viral RNA) cloned into an *E. coli* plasmid. Normal tobacco plants were infected with the *in vitro* transcribed RNA (NB the complete virus particle with coat protein is not needed for mechanical infection). The recombinant TMV RNA was infectious and a high yield of chimeric virus particles was obtained from infected plants.

The recombinant viruses were shown to present the malarial epitope on the surface and there is a good prospect that they will be immunogenic. This opens up the prospect of producing malaria vaccines at low cost in areas of the world where it is needed. Since the TMV viral rods are very stable at ambient temperature, there are no costs or problems of refrigerated storage of the vaccine.

Example 2. Bulk industrial enzyme production: α-amylase

α-Amylases are widely used in the food and detergent industries. The enzyme hydrolyses the α-1,4-glycosidic linkages in starch and is used in starch processing, in the brewing industry for the production of low-calorie beer, in the baking industry for increasing bread volume, in the wine industry for clarification and in biological detergents. It is currently produced by bacterial fermentation from *Bacillus licheniformis*. The enzyme produced by this bacterium is heat stable and active over a broad pH range. These properties are particularly useful in industrial processes since heat-gelatinized starch is a better substrate for the enzyme than native starch.

Two chimeric α-amylase genes were constructed using the *B. licheniformis* coding sequence. Both had the CaMV constitutive 35S promoter and the *nos* polyadenylation signal but differed in the N-terminal signal sequence used. One construct used the signal sequence from a tobacco extracellular pathogen-responsive protein (PR-S) and the other used the bacterial export signal from the α-amylase gene itself. Interestingly, both proteins were exported from the cells of transgenic tobacco plants with equal efficiency so that α-amylase accumulated in the extracellular spaces. The molecular weight of the transgenic plant enzymes was higher than that of the bacterium (64×10^3 M_r versus 55.2×10^3 M_r). This difference is due to post-translational glycosylation of the enzyme by the plant. There are six potential asparagine glycosylation sites on the bacterial α-amylase sequence and digestion of the recombinant plant enzyme with *endo-N*-acetylglucosaminidase H, which removes asparagine-linked oligosaccharides from proteins, reduces the molecular weight of the transgenic plant protein to $55 \times 10^3\ M_r$, which confirms the glycosylation. Despite glycosylation the plant enzyme was able to degrade starch at 95–100°C.

The transgenic tobacco plants were phenotypically the same as the parent cultivar and, as expected from the extracellular compartmentalization of the α-amylase, the starch content of the tobacco leaves was not altered in the transgenic plants. Studies on the plant-produced enzyme showed that the major degradation products from starch were the same as those obtained with the *Bacillus* enzyme and this suggests that the plant enzyme can substitute for the bacterial one in industrial processes.

16.3 The plant biotechnology debate: the Frankenstein factor

Despite the acknowledged potential benefits of plant biotechnology to both farmers and consumers, a number of concerns associated with its use are being debated. In this section the nature of these concerns and some relevant aspects of the technology and its implementation will be presented, without intended participation in the debate itself.

In 1994 the UK Biotechnology and Biological Sciences Research Council (BBSRC) funded a consensus conference organized in London by the Science Museum. This conference was modelled upon a similar conference system developed in Denmark. Sixteen lay volunteers set the agenda for the conference by deciding upon the key questions that would be addressed. They also chose the expert witnesses who were called to attend, conducted the questioning and wrote a final report, which was published and widely distributed to various bodies and to plant molecular scientists. This was a pioneering venture to investigate public opinion about plant biotechnology.

Seven questions were formulated:

Question 1. What are the key benefits and/or risks of modern plant biotechnology?

Question 2. What possible impact could plant biotechnology have on the consumer?

Question 3. What possible impact could plant biotechnology have on the environment?

Question 4. What moral problems are raised by plant biotechnology?

Question 5. Why are patenting and intellectual property rights such a feature of plant biotechnology?

Question 6. How can we ensure that plant biotechnology benefits rather than harms the developing world now and in the future?

Question 7. What are the prospects for effective regulation of plant biotechnology?

These questions essentially encompass all the topics that have been more widely debated.

Apart from the ethical concerns that biotechnology is not 'natural' and is producing what has been called 'Frankenfood', there are also problems for some groups in the manipulation of animal genes in plants. The UK Consensus Conference felt that proper labelling was important for these issues.

Several potential risks associated with plant biotechnology can be grouped as follows:

1. Environmental risks:
    (a) transfer of genes to other plants (e.g. creating new weeds), with consequential detrimental effects on the environment;
    (b) selection or production of new pests or diseases;
    (c) increase in environmentally detrimental practices (e.g. monoculture and chemical use).
2. Health hazards from new food products or pharmaceutical products.
3. Risks of undermining economies in the developing world.

Environmental risks

Definitive answers to questions concerning the potential environmental risks do not exist. In the UK an independent body, the Advisory Committee on the Release to the Environment (ACRE), is responsible for monitoring the safety of all transgenic plants grown in the field. The equivalent body in the USA is the Environmental Protection Agency (EPA), which issues licences for the growth of transgenic plants in both field tests and commercial production. During the late 1980s 10 UK companies with interests in plant biotechnology, together with the Department of Trade and Industry and the Agricultural and Food Research Council, inaugurated a study to evaluate the environmental impact of releasing transgenic plants for growth in fields. The study was called Planned Release of Selected and Modified Organisms (PROSAMO). The study ended in 1993 and concluded that the release of transgenic plants was not an unacceptable threat to the environment. In an experimental study, which was regulated by another body, the Intentional Release Sub-committee of the Advisory Committee on Genetic Modifications, it was found that there was no evidence of cross-pollination between transgenic potato plants and the related weeds black nightshade and woody nightshade. The method of risk assessment used in this study was to monitor the transfer of the bacterial kanamycin resistance gene from transgenic potato to the weed species.

The long-term selection or production of new pests and diseases is more difficult to assess but, for example, strategies to delay the evolution of pests resistant to *Bt* toxins have been developed. These are based upon the use of patchworks of transgenic and non-transgenic fields, which may give good protection to both, provided the biology of the pest is understood and used to devise the patchwork pattern.

In the light of some of the early claims for a biotechnological revolution, it is perhaps disappointing that the major trait currently manipulated in plant biotechnology is herbicide resistance. This follows from the fact that the technology (particularly in the USA) is largely funded by industry and it is not surprising that projects that are going to provide a profit on the invested development funds will be chosen over other possibly more altruistic projects. The chemical companies claim that the development of transgenic herbicide-tolerant crops will allow the more 'environment friendly' herbicides to be used and this a positive environmental result from this work.

Health hazards

The possible health hazards of new food products in the UK are the responsibility of the Advisory Committee on Novel Foods and Processes, which is an advisory committee of the Ministry of Agriculture, Fisheries and Food. In the USA it is the responsibility of the Food and Drug Administration.

The first transgenic plant to go on the market was the Flavr Savr tomato produced by Calgene (Davis, California, USA). In the USA, the tomato can be sold to consumers either processed in soups and ketchups or as fresh fruit but in the UK the fresh, uncooked, produce has not been given a licence. The reason for the limitation in the UK is not the transgenic nature of the cultivar nor the antisense polygalacturonase construct used, but the fact that the transformation technique involves the use of a bacterial antibiotic resistance (kanamycin) gene that is co-transformed with the antisense polygalacturonase construct in order to select transformed cells. This was considered to be a potential hazard if somehow acquired by pathogenic or potentially pathogenic microorganisms.

The problem with differences in national licensing authorities is well illustrated by transgenic tomato. Because tomato seeds are resistant to damage by digestive enzymes, they can pass through the gut and subsequently germinate. A possible scenario is that someone eats a Flavr Savr tomato sandwich in a USA airport and travels to the UK carrying the transgenic seed in their intestine. The seed can then escape, germinate and grow in the UK as an innocuous 'escape' tomato plant with no controls over its use.

Impact on the developing world

The possible effects of plant biotechnology on the developing world is a complicated area of debate. There are three main concerns.

1. Funding is limited for expensive molecular research projects on tropical crops important in the developing world but which have no market value in the developed world. Cassava is an example of this problem. It is the third most important crop grown in the tropics, estimated to be the staple food of some 500 million people. In Africa it is particularly important and is arguably the most important crop for the rural poor grown in tropical Africa. It is known as an orphan crop because of the very low international budget for research. Recently an international network (The Cassava Biotechnology Network) has been funded by The Netherlands; this organization aims to promote and coordinate molecular cassava projects in both the developed and the developing world.

Other organizations, such as the International Service for the Acquisition of Agri-Biotech Applications based in Ithaca, New York, USA, have also been set up to address this problem in other crops.

2. There is also a concern that products currently produced in the developing world will be the basis of plant biotechnology projects, resulting in the production of these compounds in the developed world and thus removing vital international revenue from fragile economies. The production of lauric acid in transgenic canola (see section 15.4) that can be grown in North America is an example of this problem because currently lauric acid is imported primarily from South-East Asia in palm kernel and coconut oil.

3. The third concern affecting the developing world also has much wider implications. The problem arises from the very broad patents that have been awarded in plant biotechnology, effectively patenting all transgenic manipulations of a particular crop. In 1992 the USA Patent and Trademark Office (PTO) awarded a patent to Agracetus Inc. of Middleton, Wisconsin, USA for the rights of all genetically engineered cotton, regardless of the method of transformation or the transgene used. In March 1994 the European Patent Office (EPO) granted broad rights to the same company (Agracetus Inc.) for all forms of genetically engineered soybean. Biotechnology companies expect to use patents to protect their products and the substantial investment incurred in the development of those products. In the period from the early 1980s to 1995 the PTO had awarded 112 patents for genetically engineered plants and for molecular techniques for manipulating plants; however the very broad patents awarded to Agracetus Inc. have raised alarm both among research scientists and among other plant biotechnology companies. Both of the Agracetus Inc. patents have been challenged and may in time be revoked or revised. However, they have highlighted the issue of patents and the extent to which they may either hinder or promote progress in plant biotechnology. The issue of patenting also impinges upon the relationship between public funded research in plant molecular genetics and private sector research in the development of applications of these innovations in plant biotechnology. Many of the innovations and the genes used in plant biotechnology have come from public sector research and these broad patents threaten the open development of the science.

Major Learning Objectives for Chapter 16

1. Knowledge of the scope of the applications of plant biotechnology.
2. Understand the nature of the risks associated with the production of transgenic plants with reference to:
   (a) environmental risk,
   (b) health hazards,
   (c) the developing world.

Further reading

SCHMIDT, K. (1995). Whatever happened to the gene revolution? *New Scientist*, **145**, 21–25.
A broadly based article that questions the benefits of plant biotechnology.

STONE, R. (1995). Sweeping patents put biotech companies on the warpath. *Science*, **268**, 656–658.
This short article debates the issues associated with patenting in plant biotechnology.

The following three references describe the biotechnology of three different transgenic plant projects and illustrate the potentials and the constraints of the technology.

MOFFAT, A.S. (1995). Exploring transgenic plants as a new vaccine source. *Science*, **268**, 658–660.

PEN, J., SIJOMS, P.C., VAN OOIJEN, A.J.J. and HOEKEMA, A. (1993). Protein production in transgenic crops: analysis of plant molecular farming. In Hiatt, A. (ed.) *Transgenic plants*, pp. 239–251. Marcel Dekker, New York.

POIRIER, Y., NAWRATH, C. and SOMERVILLE, C. (1995). Production of polyhydroxyalkanoates, a family of biodegradable plastics and elastomers, in bacteria and plants. *Bio/Technology*, **13**, 142–150.

Ceres is a journal published by the United Nations Food and Agriculture Organization and the May–June 1995 issue contained a special centrepiece on the impact of plant and animal biotechnology on the developing world. The following three articles come from this issue.

GOLDBERG, R. (1995). Pause at the amber light. *Ceres*, **153**, 21–25.

JONAS, D. and KAFERSTEIN, F. (1995). Manning Dinner's Defences. *Ceres*, **153**, 26–28.

JUMA, C. and MUGABE, J. (1995). Get up, stand up, keep up. *Ceres*, **153**, 34–40.

The September 1995 issue of *Trends in Biotechnology* (vol.13, no.9) is a special issue devoted to plant biotechnology and contains the following four articles.

GODDIJN, O.J.M. and PEN, J. (1995). Plants as bioreactors. *Trends in Biotechnology*, **13**, 379–387.

MASON, H.S. and ARNTZEN, C.J. (1995). Transgenic plants as vaccine production systems. *Trends in Biotechnology*, **13**, 388–392.

DALE, P.J. (1995). R&D regulation and field trialling of transgenic crops. *Trends in Biotechnology*, **13**, 398–403.

TOENNIESSEN, G.H. (1995). Plant biotechnology and developing countries. *Trends in Biotechnology*, **13**, 404–409.

Index

Page numbers in italic refer to figures.